The Art of Creative Math

Eighteen Ideas That Make Hard Problems Easy

Dr Hui Quan

ASCENT MATH PRESS

The Art of Creative Math: Eighteen Ideas That Make Hard Problems Easy.

Published by Ascent Math Press.

The diagrams in this book were produced in Ti*k*Z; the text is set in TeX Gyre Pagella and Heros.

First edition.

A companion volume, *The Art of Creative Math: Complete Solutions*, contains full worked solutions to every problem in the practice ladders. In this book, the answer keys give brief NUDGES only — enough to confirm a result or unstick a start, never the whole road.

A note before you begin

This is not a problem book that happens to have explanations. It is a book of *ideas* that happens to have problems — eighteen of the most beautiful ideas in competition mathematics, each strong enough to crack open a hundred problems that look, at first, like they have nothing in common.

Most contest books are organized like toolboxes: a drawer of counting problems, a drawer of number theory, go practice. That is a fine way to drill and a poor way to fall in love. It hides the secret that makes mathematics worth doing: that the same few ideas keep reappearing in disguise, and that learning to *recognize* them is the real skill. A double-counting argument and a probability calculation can be the same thought in different clothes; a recurrence, a bijection, and a generating function, three windows onto one sequence of numbers. See those connections and problems stop being a wall of unrelated puzzles and become a landscape you can navigate.

Threaded through the book are SEEN BEFORE? boxes, pointing to the moments when one chapter's idea quietly resurfaces in another. Follow them; they are the connective tissue, and they are the point. The Catalan numbers appear in three chapters by three different methods; the humble fact that a square is never negative underlies an entire chapter on inequalities; double-counting, the very first idea in the book, turns out to hide inside Pascal's triangle, inside graph theory, and inside probability.

Two old stories say everything about the spirit of this book. The first: a schoolteacher, hoping for a quiet hour, asks the class to add the whole numbers from 1 to 100. A boy of perhaps eight or nine — Carl Friedrich Gauss — sets down his slate almost at once with the single number 5050. He had not added a hundred numbers. He had *noticed* that 1 and 100, 2 and 99, 3 and 98, and so on, each pair to the same 101, and that there are fifty such pairs: $50 \times 101 = 5050$. The arithmetic everyone else was grinding through, he had dissolved with one act of seeing. That is creative mathematics in a sentence: not more effort, but a better way of looking, so that the hard problem quietly becomes an easy one. (You will meet Gauss's pairing trick again, in its grown-up forms, in the chapters on telescoping and on counting.)

The second story is older. Archimedes, asked to judge whether a crown was pure gold without melting it down, is said to have found the answer as he settled into a bath: watching the water rise, he saw that a submerged body displaces its own volume, so the crown's density could be measured after all. He leapt up shouting *Eureka!* — "I have found it!" The word now names a feeling every solver knows: the click when a tangle suddenly comes clear, often from an angle nobody pointed out. This book is, more than anything, an attempt to engineer those clicks on purpose. Watch for your own small *eurekas*; they are the sound of an idea taking hold.

In that spirit, a companion travels with you: **Curious Carl** — a bright, hard-working student taught, as nearly all of us are, that mathematics is a matter of *effort*: apply the algorithm, push through the arithmetic, and grind until the answer gives way. So his first instinct is always to work *harder*. He turns up in each chapter to reach for the calculator, list every case, crank the formula — and to be delightfully wrong before he is right. Watching him toil, then stumble onto the one clean idea that makes all the toil unnecessary, is the point. Carl is not learning to beat a test. He is learning to be *lazy in the most brilliant way there is*: to refuse to grind when there is a way to *see*.

Curious Carl

over-prepared,
over-computing,
about to learn
a better move.

So each chapter is built around *one idea* and follows the same rhythm. A WONDER THIS box opens with a question you can start attacking in one sitting but cannot cleanly finish without the chapter's idea. Then the idea itself, in plain language, reasoned rather than asserted. WATCH THIS boxes model the move in action; TRAPDOOR boxes flag the mistake almost everyone makes; and a SHOWPIECE shows the idea at full power. Every chapter ends with a one-page recap and a *practice ladder* that climbs from warm-ups to genuine challenges.

A word on how to read. Each chapter is meant for one focused sitting, roughly forty-five minutes to an hour, and the WONDER THIS box that opens it is not decoration; it is the starting gun. Sit with that puzzle for at least ten minutes, pencil moving, *before* you read on. Read passively and you will follow every step and keep almost none of it. Treat the answer-key nudges as hints, not solutions, and reach for the companion volume's full solutions only when you are genuinely stuck or want to check your work. Reading the solution before attempting the problem is like reading the last page of a mystery first: efficient, and ruinous. And read the prose between the boxes — the boxes are the skeleton, but the reasoning that connects them is where the understanding actually lives.

And a word on being stuck. Being stuck is not a malfunction; it is the sound of your brain building something new. Every mathematician who ever lived, Gauss included, has spent most of their working hours stuck. When you are, you have three reliable friends, and Curious Carl will reach for all of them: *small cases* (try $n = 1, 2, 3$ and watch for a pattern), *a different representation* (draw the picture, write the equation, tell the story), and *walking away* (your back-of-mind keeps working while you do something else). Competition culture prizes speed; this book does not. The struggle is not the obstacle in front of the learning — it *is* the learning.

You do not need to read the chapters in order, though the cross-references are richest if you do. You do not need to be preparing for any particular competition; nothing here is tied to one. You need only a willingness to sit with a good question long enough to be surprised by it. That willingness, more than any technique in these pages, is the real art of creative math.

How to climb a practice ladder

Every chapter ends with a *practice ladder*: a sequence of problems sorted not by topic but by what they ask of you. The rungs are labeled, and the labels mean something. Use them to decide where to start and how hard to push.

THE PRACTICE LADDER *climb at your own pace*

LAUNCH Warm-ups. One step, using the idea in its plainest form. If these don't flow, reread the chapter before climbing.

CORE BUILDER The heart of the chapter. These are the problems you should be able to do cold once you've absorbed the idea — the ones worth mastering before moving on.

STRETCH Transfer. The idea applied on less familiar ground, where you must first *recognize* that it's the right tool.

THE FLASH No clock. Each of these is built to crack open in a single flash of insight, the moment you find the right angle — so if you're grinding, you haven't found it yet. Keep hunting for the view that makes it easy.

PROOF LAB Write it out in full sentences. Proving the chapter's central results yourself is the surest test that you truly own the idea.

REFLECTION No nudge, no single answer. A prompt to put words to what you just learned — because the student who can explain *why* a method works is the one who will reach for it again.

The answer key after each ladder gives brief NUDGES: enough to check your result or restart a stalled attempt, never the full path. Full worked solutions are in the companion volume. Resist it until you've fought.

Contents

The Creative Moves Map

Eighteen ideas — and the threads that tie each to the others.

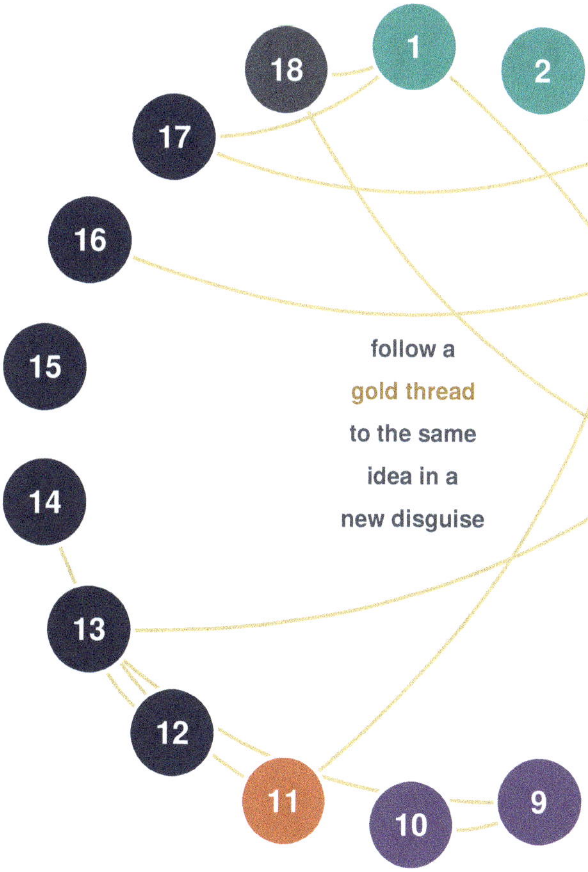

1 Double Counting
2 Too Many Pigeons
3 What Can't Be Undone
4 Count Something Else
5 Solve It Smaller
6 The Great Cancellation
7 The Triangle That Runs Everything
8 A Clothesline for Numbers
9 Only the Remainder Matters
10 When Giants Collapse
11 The Honeybee's Craft
12 The Back Door to the Imaginary
13 Tools From the Imaginary Box
14 The Picture Behind the Formula
15 The Lever and the Shadow
16 Counting Past the Mirror
17 Dots, Lines, and Lonely Hearts
18 The Unpecked Chicken

Seeing It Differently
Translating the Problem
The Arithmetic of the Infinite
The Optimum Is Balance
Changing the Universe
The Grand Average

CHAPTER ONE

1

Double Counting

The art of counting the same thing twice — on purpose — and getting paid in identities.

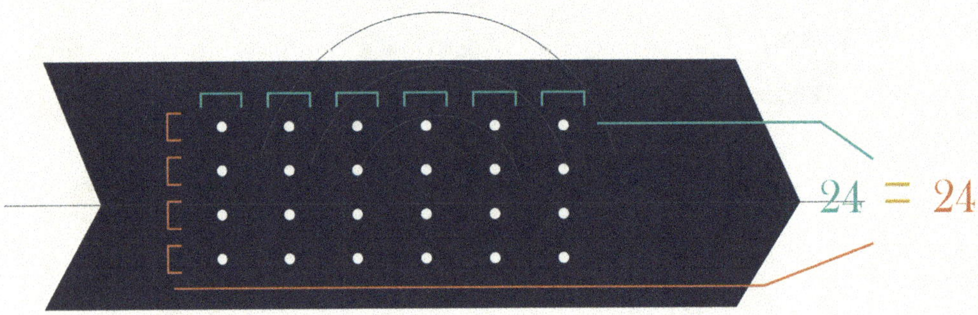

Double counting: one set, two stories.

Here is a question you can attack right now, no theory required.

> **WONDER THIS**
>
> Write down every four-digit number that uses each of the digits $1, 2, 3, 4$ exactly once. There are 24 of them: 1234, 1243, 1324, all the way up to 4321.
>
> **Add them all up.** What is the total?

You could, in principle, list all twenty-four numbers and grind out the sum by hand. Please don't. That is the path of suffering, and it is also the path of arithmetic mistakes. There is a better way, and it is the seed of this entire chapter. Before you calculate, look again: the first escape is not a trick but a posture. Change *what* you look at, and the labor can vanish — a wall becomes a door.

CURIOUS CARL • FIRST GUESS

Carl: "A handshake party? I'll walk the room, ask each person how many hands *they* shook, and total it up. Bookkeeping — what could go wrong?"

The catch: that adds every handshake in *twice*, Carl, once from each side — but far from a mistake, that doubling is the whole trick we're about to lean on.

Instead of adding the numbers *row by row*, add them up *column by column*: ones place, tens place, hundreds place, thousands place. Ask a sharper question: across all 24 numbers, how many times does the digit 3 land in the tens column? If we glue 3 into the tens slot, the other three digits $\{1, 2, 4\}$ can fill the remaining slots in $3! = 6$ ways. So 3 shows up in the tens column exactly 6 times. And there was nothing special about the 3 or about the tens. *Every* digit lands in *every* column exactly 6 times.

That single observation does all the work. The ones column contributes $6 \times (1 + 2 + 3 + 4) \times 1$ to the grand total; the tens column contributes $6 \times (1 + 2 + 3 + 4) \times 10$; and so on up the place values. Factor out the common piece:

$$\text{Total} = 6 \times (1 + 2 + 3 + 4) \times (1 + 10 + 100 + 1000) = 6 \times 10 \times 1111 = \boxed{66{,}660}.$$

No grinding. We computed the same sum two different ways (once imagined row by row, once organized column by column) and the column story was so much easier that the answer practically fell into our laps. Underneath the arithmetic, one move is doing all the work: **if you count the same collection of things in two different ways, the two answers have to agree.** So pick the easy way, and let the hard way come along for free.

THE MOVE

Pick one set of objects. Count it one way; call the answer L. Count the *same* set a second, cleverer way; call the answer R. Conclude $L = R$. The skill is never the arithmetic — it is *choosing the set* and *choosing the two stories* so that both counts are easy and the equation they force is the one you wanted.

It sounds almost insultingly obvious: of course a number equals itself. And yet this idea will hand you identities that look like they should cost a page of algebra, it will explain a famous fact about graphs, and (in the showpiece of this chapter) it will tell you why there are exactly *five* perfectly symmetric solids in the universe and not a sixth. **Double counting** is mathematics catching the same fact on camera from two different angles. The footage has to match. And when it matches, an identity is already in the frame, whether you went looking for it or not.

The cleanest case: rows versus columns

The digit trick was, once translated, a table. The rows were the 24 numbers; the columns were the four place-values; and we counted the total by sweeping columns instead of rows. That picture (*a grid you can total two ways*) is the friendliest form of double counting, and it deserves to be made explicit, because once you see it you will see it everywhere.

Whenever you can arrange your objects into a table and drop a mark in some cells, you have set up two counts for free: **add up the rows**, or **add up the columns**. The total number of marks does not care which way you sweep. So the two sweeps are equal. Here is the most famous example in all of graph theory, and it shows how a one-sentence idea hardens into a theorem.

WATCH THIS — **#1 The Handshake Lemma**

Claim. At any party, the number of people who shook an *odd* number of hands is even.

The set we count. Make a table: people down the side, handshakes across the top. Put a dot in a cell when that person was one of the two hands in that handshake. We will count the dots.

	h_1	h_2	h_3	h_4
Ana	•	•		
Bo	•		•	
Cy		•	•	•
Di				•

row sum = your degree

column sum = 2

Count by rows. Ana's row holds as many dots as hands Ana shook — her *degree*. So the row totals add up to $\sum_{\text{people}}(\text{degree})$.
Count by columns. Every handshake involves exactly two hands, so every column holds exactly *two* dots. With H handshakes, that is $2H$ dots.
The footage matches.

$$\sum_{\text{people}} \deg(\text{person}) = 2H.$$

The right side is even. So the left side — the sum of everybody's handshake count — is even too. The even-degree people contribute an even chunk; whatever is left over, contributed by the odd-degree people, must *also* be even. A sum of odd numbers is even only when there is an even number of them. Done. ■

Trigger A quantity that two different sweeps of a table can both compute (here, dots = handshake-incidences).

Do first Build the people-by-handshakes table and ask what a row total and a column total each mean.

Trapdoor Counting handshakes as if each contributes one dot. It contributes *two* — one per hand. That factor of 2 is the entire punchline.

Upgrade Replace "2 hands per handshake" with "k slots per object" and you get the general incidence identity $\sum \deg = k \cdot (\#\text{objects})$.

The seductive error here is to write $\sum \deg = H$ instead of $2H$, reasoning that "there are H handshakes." But we are not counting handshakes. We are counting *dots*, and each handshake spends two of them. This is worth burning into memory, because it is the way a perfectly good double count most often self-destructs: whenever you count incidences, stop and ask how many times each object gets counted before you commit to a number.

△ TRAPDOOR

Off-by-a-factor is the classic double-counting death. "Two hands per handshake," "three edges per triangle," "each pair counted from both ends" — these little multipliers are exactly where the count quietly goes wrong. Before you equate your two stories, say out loud how many times each object is counted in each story.

CURIOUS CARL • TRAPDOOR!

Carl counts the ways two of five friends can shake hands: "Five people, and each has four others to greet — so $5 \times 4 = 20$ handshakes!" He writes it down, pleased.

One lonely handshake pipes up from the page: *"You counted me from Ana's side and again from Bo's side — I'm one handshake, not two."* Right, Carl: every pair got tallied from both ends, so the true count is $20/2 = 10$. Divide when each object is counted twice — and *don't* divide when it isn't.

There is a reusable instinct hiding in all of this, and I want you to hear it as an actual alarm bell. The moment a table appears (or the moment you can *imagine* one), your brain should ping: *rows or columns?* If the rows are easy, sweep rows. If the columns are easy, sweep columns. If *both* are easy, congratulations: you are about ten seconds from proving something, and you may feel appropriately smug. Most students can follow a double-counting proof and still never produce one, and the reason is almost always that nobody trained them to ask the second question. School trains you to compute the quantity in front of you. Double counting asks a different question. *What is a second thing this same quantity also counts?* That question is the entire game.

And the two sweeps do not only prove *identities*. They also *audit*, forcing a bound when one count is a rate and the other is a cap.

WATCH THIS — #2 Double counting as an audit

Claim. Twenty students each solve exactly 6 problems on an exam, and each problem is solved by at most 5 students. Then the exam has at least 24 problems.

Count the solved pairs. Tally the pairs (student, a problem they solved) two ways. Sweeping by students, each of the 20 solves 6 problems, giving $20 \times 6 = 120$ pairs. Sweeping by problems, each problem is solved by at most 5 students, so it contributes at most 5 pairs; with P problems that is at most $5P$. The two sweeps count the same pairs, so

$$120 \le 5P \quad \Longrightarrow \quad P \ge 24.$$

No exam with fewer than 24 problems can meet the conditions. ■

Here double counting proves nothing about a formula — it *bounds* a quantity by pitting a per-student rate against a per-problem cap. *When the first count feels natural, ask for the second camera angle — and count the shadow, not the thing.*

The committee with a leader

Rows and columns are double counting wearing a name tag. The deeper version drops the grid and just tells *two stories about the same scenario*. Here is the single most useful storytelling template you will ever learn. It cracks a whole family of identities that look algebraically nasty.

WATCH THIS

#3 The gloriously unnecessary Team Captain

Claim. $\sum_{k=0}^{n} k\binom{n}{k} = n\,2^{n-1}$.

Stare at the left side and it dares you to expand it. Don't take the bait. Instead, *tell a story the left side is counting.*

The scenario. From n students, form a committee of any size, and then hand exactly one member the gloriously unnecessary title of *Team Captain*. We count the pairs (committee, captain) two ways.

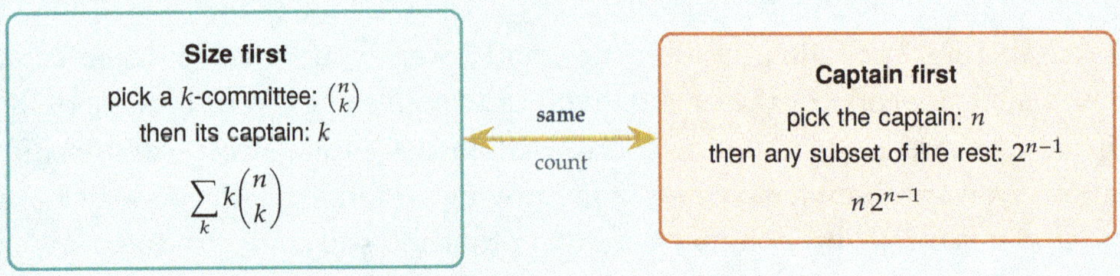

Size first. Choose the committee, then crown someone: $\sum_k \binom{n}{k} \cdot k$.
Captain first. Crown the captain (n choices), then let each of the remaining $n-1$ students independently be in or out (2^{n-1} ways): $n\,2^{n-1}$.
Same universe, two camera angles. The binomial identity practically introduces itself:

$$\sum_{k=0}^{n} k\binom{n}{k} = n\,2^{n-1}. \blacksquare$$

THE TURN

from grind to insight

The obvious way to prove an identity like this is to roll up your sleeves and *expand* — rewrite $k\binom{n}{k}$ as $n\binom{n-1}{k-1}$, push factorials around, and grind the algebra until the two sides finally agree. It works, but it is joyless, and it never tells you *why* the identity is true. **The creative turn** is to refuse the algebra entirely and instead describe *one* collection — committees with a captain — then count that same collection two different ways. The two counts *must* agree, so the identity falls out with no manipulation at all. That is the leap of this chapter: don't transform the equation — find the set both sides are secretly counting, and let consistency do the work.

Trigger An algebraic sum with a stray factor of k (or $\binom{n}{k}$ times something) multiplying a binomial coefficient — that k is begging to become "choose a distinguished person."

Do first Invent a scenario the sum counts: a committee plus one special role. Count it by size, then by role.

Trapdoor Trying to expand $k\binom{n}{k}$ algebraically. You *can* (it equals $n\binom{n-1}{k-1}$), but the story is faster and tells you *why*.

Upgrade Two special roles (captain *and* secretary) gives $\sum k^2\binom{n}{k}$-type identities; the same staging handles them.

Notice what just happened: we never did any algebra. We described a situation, counted it from the left, counted it from the right, and the identity fell out. That is the signature feeling of a great double count. It feels less like calculation and more like catching the universe being consistent with itself. Keep that sensation in mind, because we are about to point it at something far bigger than a binomial coefficient.

Two rooms: a first taste of Vandermonde

Here is a double count whose answer looks, at first, like it has no business being so clean.

WATCH THIS **#4 A committee from two rooms**

Claim. $\sum_{k=0}^{n}\binom{n}{k}^2 = \binom{2n}{n}$.

The scenario. A club has n women and n men — two rooms of n people each — and wants to choose a committee of exactly n people from the $2n$ total. We count the number of such committees two ways.

Count one (ignore the rooms). A committee of n chosen from all $2n$ people: simply $\binom{2n}{n}$.

Count two (split by the rooms). Suppose the committee takes k women; then it must take the remaining $n-k$ members from the men. The women can be chosen in $\binom{n}{k}$ ways and the men in $\binom{n}{n-k}$ ways. Since $\binom{n}{n-k} = \binom{n}{k}$, that is $\binom{n}{k}^2$ ways for each k. Summing over all possible k:

$$\sum_{k=0}^{n}\binom{n}{k}\binom{n}{n-k} = \sum_{k=0}^{n}\binom{n}{k}^2.$$

The two counts count the same committees, so they are equal. **The sum of the squares of row n of Pascal's triangle is the central entry of row $2n$ — the number $\binom{2n}{n}$.** (Check at $n = 2$: $1^2 + 2^2 + 1^2 = 6 = \binom{4}{2}$.) ■

The squares appeared for a reason (the mirror symmetry $\binom{n}{n-k} = \binom{n}{k}$), not by accident. This is the simplest case of *Vandermonde's identity*, and you will meet its full form, $\sum_k \binom{m}{k}\binom{n}{r-k} = \binom{m+n}{r}$, proved by the very same two-rooms argument, in **Chapter 7**.

The showpiece: why exactly five Platonic solids?

A *Platonic solid* is the most democratic shape imaginable: every face is the same regular polygon, and every vertex looks exactly like every other. The ancient Greeks knew five of them (the tetrahedron, cube, octahedron, dodecahedron, and icosahedron) and suspected there were no others. But "we looked really hard and couldn't find a sixth" is not a proof. Double counting turns that hunch into certainty.

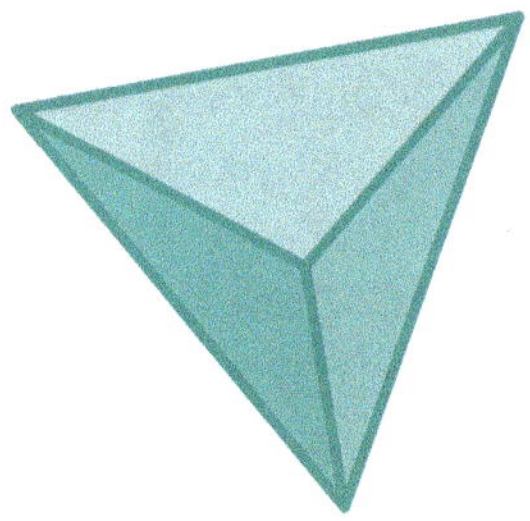

tetrahedron

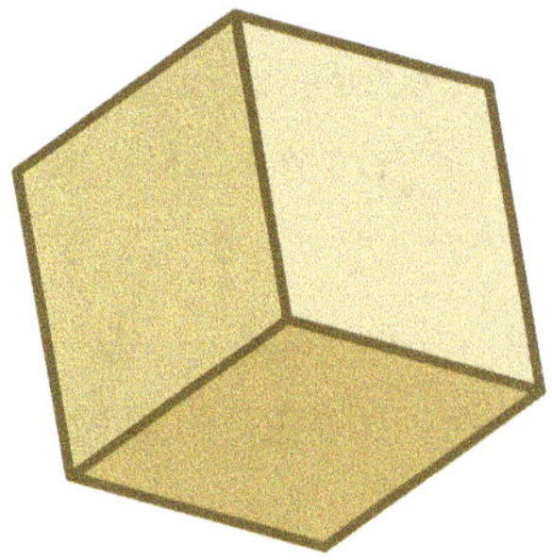

cube

octahedron

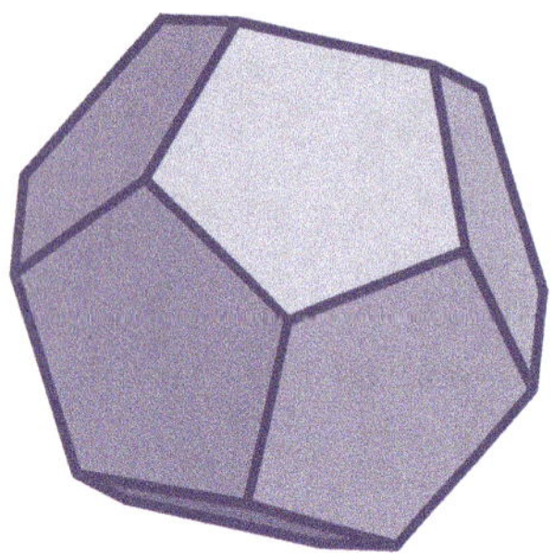

dodecahedron

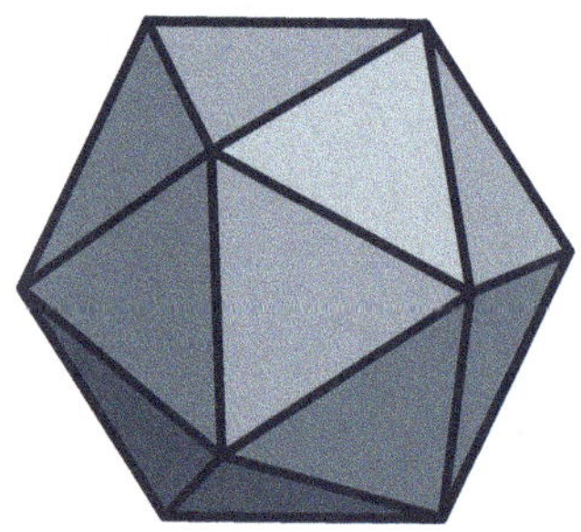

icosahedron

WATCH THIS **#5 Five, and not one more**

The set we count. The *edges* of a Platonic solid — counted three different ways, through faces, through vertices, and through Euler's bookkeeping.

Let every face be a regular p-gon, and let exactly q faces (equivalently, q edges) meet at each vertex. Write V, E, F for the number of vertices, edges, and faces. Now we double count edges twice over.

Edges through faces. Each face is bordered by p edges. That accounts for pF edge-borders — but every edge sits between exactly *two* faces, so we've counted each edge twice:

$$pF = 2E.$$

Edges through vertices. Each vertex is the meeting point of q edges. That accounts for qV edge-ends — but every edge has exactly *two* ends, so again we've double counted:

$$qV = 2E.$$

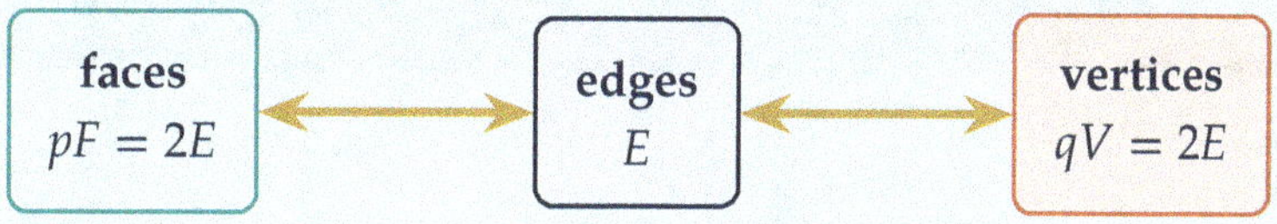

So $F = \dfrac{2E}{p}$ and $V = \dfrac{2E}{q}$. Now feed these into **Euler's formula**, the great accounting identity that every convex polyhedron obeys:

$$V - E + F = 2 \quad\Longrightarrow\quad \frac{2E}{q} - E + \frac{2E}{p} = 2.$$

Divide through by $2E$ (which is safely positive) and rearrange:

$$\frac{1}{p} + \frac{1}{q} = \frac{1}{2} + \frac{1}{E} > \frac{1}{2}.$$

It all comes down to one little inequality. A polygon needs at least 3 sides, so $p \geq 3$; at least 3 faces must meet at a corner to make a solid angle, so $q \geq 3$. We need $\frac{1}{p} + \frac{1}{q} > \frac{1}{2}$ with both integers at least 3. Test them:

p	q	$\frac{1}{p} + \frac{1}{q}$	solid
3	3	$\frac{2}{3} > \frac{1}{2}$ ✓	tetrahedron
3	4	$\frac{7}{12} > \frac{1}{2}$ ✓	octahedron
3	5	$\frac{8}{15} > \frac{1}{2}$ ✓	icosahedron
4	3	$\frac{7}{12} > \frac{1}{2}$ ✓	cube
5	3	$\frac{8}{15} > \frac{1}{2}$ ✓	dodecahedron
4	4	$\frac{1}{2} \not> \frac{1}{2}$ ✗	— (flat tiling)
3	6	$\frac{1}{2} \not> \frac{1}{2}$ ✗	— (flat tiling)

Every larger pair only makes $\frac{1}{p} + \frac{1}{q}$ smaller, so nothing beyond the table can ever clear the bar. **Exactly five solutions. Exactly five Platonic solids.** ■

Sit with that for a second.

There is no sixth Platonic solid — not because nobody has been clever enough to discover it, but because the universe will not permit it. It is one of the most satisfying facts in mathematics, and the same little count rules it out forever. We learned the deepest fact about three-dimensional symmetry by doing nothing fancier than counting edges, and then counting them again, and refusing to let the two counts disagree. The cases that *just* miss — $p = q = 4$ (squares, four to a corner) and $p = 3, q = 6$ (triangles, six to a corner) — are not failures; they are the flat tilings of the plane, the square grid and the triangular grid, sitting exactly on the boundary $\frac{1}{p} + \frac{1}{q} = \frac{1}{2}$. The same inequality that births the solids also explains your bathroom floor.

Solid	Face p-gon	q / vertex	F	E	V	$V - E + F$
Tetrahedron	3	3	4	6	4	$= 2$
Cube	4	3	6	12	8	$= 2$
Octahedron	3	4	8	12	6	$= 2$
Dodecahedron	5	3	12	30	20	$= 2$
Icosahedron	3	5	20	30	12	$= 2$

$$pF = 2E = qV \qquad \frac{1}{p} + \frac{1}{q} > \frac{1}{2}$$

The five Platonic solids: double counting faces, edges and vertices forces $1/p + 1/q > 1/2$, which has exactly five solutions.

SEEN BEFORE?

Euler's formula $V - E + F = 2$ is itself a double-counting theorem in disguise, and you'll meet it again in **Chapter 17** (Graph Theory), where we prove it by building a polyhedron one edge at a time. The trick of counting edges "through faces" and "through vertices" is the same incidence-counting move as the handshake lemma — just wearing a much more impressive hat.

TRY TO BREAK IT

The handshake count proved that $\sum \deg = 2E$ is always *even*. Now try to build a party that defies it: seat guests so that an *odd* number of them each shook an *odd* number of hands. Sketch a few and you'll find you never can — the odd-degree people always come in pairs. Ask yourself *why* the double count $\sum \deg = 2E$ forbids an odd number of odd-degree guests, and you'll have re-proved the Handshake Lemma with your own hands.

Payoff: a ninety-second mind-reading trick

One more, and this one comes with a story. Here is a puzzle that looks like it needs a roster, a spreadsheet, and an afternoon. Instead it falls in about ninety seconds.

WONDER THIS

A school has some clubs. Every pair of clubs shares *exactly one* student. A visiting mathematician glances at the situation and, without ever counting the clubs one by one, uncovers a single tight relationship binding the whole membership pattern.

How?

WATCH THIS #6 Counting through the shared student

The set we count. *Pairs of clubs*, each tagged with the one student they share. Let c be the number of clubs, and d_p the number of clubs that student p belongs to.

Story one — by pairs of clubs. There are $\binom{c}{2}$ pairs of clubs, and each pair contributes exactly one shared student, hence exactly one tagged pair. Total: $\binom{c}{2}$.

Story two — by students. A student p in d_p clubs is the shared student for every pair of *those* clubs — that's $\binom{d_p}{2}$ pairs. Summing over students: $\sum_p \binom{d_p}{2}$.

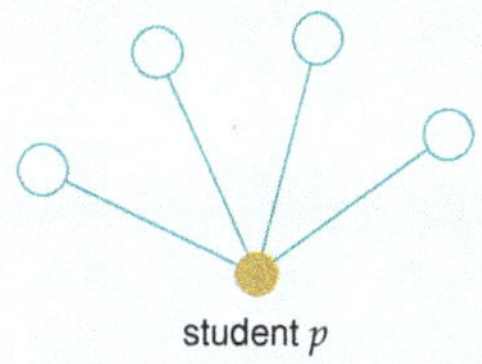

The footage matches.

$$\boxed{\sum_p \binom{d_p}{2} = \binom{c}{2}.}$$

This single identity doesn't pin down each d_p on its own — it *constrains* the whole membership pattern, tying the individual club-counts together into one equation.

That one equation is the seed of a celebrated result (Fisher's inequality) and of finite projective planes — the geometry where "every two lines meet in exactly one point" is not a paradox but a design. Our ninety-second mathematician never counted the clubs; they let the *shared students* count the clubs for them. *Show-off.* And the move that made it possible is the exact same one from the Platonic solids: when "every two things share exactly one thing," you count the pairs through the shared object.

What to try when you're stuck

When a problem smells like double counting (an identity to prove, a "how many," a configuration with a sharing rule, a sum that resists algebra), run this checklist:

1. **Name one set of objects** you could count. Just one. Resist the urge to count two different things.
2. **Find story A.** Count it the obvious way.
3. **Find story B.** Count it by a *different* organizing feature: by rows, by columns, by place value, by the largest element, by a distinguished person, by the shared object.
4. **Set the two equal.** That equation is your identity, your bound, or your impossibility.
5. **Sanity-check** on a tiny case ($n = 1, 2, 3$). If the two stories disagree, one of them is miscounting, usually by a hidden factor of two.

THE MOVE

One-page recap. Double counting proves $L = R$ by counting a single set two ways. Your toolkit: **place-value / rows vs. columns** (the digit sum, incidence tables), **committee-with-a-role** (a stray k in the sum), **count through the shared object** ("every two share one"), and **count edges three ways** (Euler, and the five Platonic solids). When stuck, name the set, tell two stories, and refuse to let the universe contradict itself.

THE NAME FOR IT

Double counting

Counting one set in two different ways forces the two answers to be equal. This is the technique known formally as *double counting* (or counting in two ways), and the equalities it produces are *combinatorial identities.*

WHERE IT LEADS Double counting is also called *Fubini's principle*: summing a two-variable quantity by rows or by columns must agree. In analysis this becomes *Fubini's theorem*, the rule that lets a double integral be evaluated in either order — the backbone of multivariable calculus, probability, and Fourier analysis.

CURIOUS CARL • EUREKA

Carl eyes his blank scratch page. "I set that whole sheet aside for the algebra. The algebra never showed up. Do I... still get to keep the answer?"

SAME BEAST, NEW COSTUME

The same "count one set two ways" beast, in costume across the book:

Handshake lemma → Pascal's identity → Swapping a double sum → Burnside → Linearity of expectation.

Each one names a single collection, counts it from two angles, and lets the footage match. Spot the shared set and the equation writes itself.

The practice ladder

THE PRACTICE LADDER

climb at your own pace

LAUNCH *Warm the engine. Full hints below.*

1. Add up every three-digit number that uses each of $1, 2, 3$ exactly once. Use the place-value trick, not brute force.

2. A graph has 7 vertices, each of degree 4. How many edges? Could every vertex instead have degree 3? Let the handshake table answer both.

CORE BUILDER *Master the central moves. Strategy notes below.*

1. Prove $1 + 2 + \cdots + (n-1) = \binom{n}{2}$ by counting the edges of the complete graph K_n two ways.

2. Prove the hockey-stick identity $\sum_{i=r}^{n} \binom{i}{r} = \binom{n+1}{r+1}$ by classifying $(r+1)$-subsets by their largest element.

3. Evaluate $\sum_{k=2}^{n} \binom{n}{k}\binom{k}{2}$ by choosing a committee, a captain, *and* a co-captain from n students — then adapt your multi-role instinct (captain *and* secretary) to attack $\sum_k k^2 \binom{n}{k}$.

STRETCH *Transfer to unfamiliar ground. Light nudges only.*

1. A convex polyhedron has 12 faces, each a pentagon or a hexagon, and exactly three faces meet at every vertex. Use double counting (edges through faces, edges through vertices) together with Euler's formula to find how many faces are pentagons.

2. Prove Vandermonde's identity $\sum_k \binom{m}{k}\binom{n}{r-k} = \binom{m+n}{r}$ by counting r-person committees drawn from m seniors and n juniors.

THE FLASH *No clock — each cracks open in one flash of insight, once you find the angle. Answers at the back.*

1. In a tournament of n players with no ties, the total number of wins equals what, and why must it equal the number of games?

2. A $3 \times 3 \times 3$ cube is built from unit cubes. Counting (small cube, exposed face) incidences two ways, how many unit faces are visible on the surface?

PROOF LAB *Write it clean, in full sentences.*

1. Let c be the number of clubs and d_p the number of clubs containing person p. Prove that in any finite collection of clubs where every two clubs share exactly one member, $\sum_p \binom{d_p}{2} = \binom{c}{2}$. Explain why this identity is a *first step* toward Fisher-type inequalities (which bound the number of clubs) — and why, on its own, it does *not* give a clubs-versus-students bound without extra uniformity hypotheses.

2. Prove there is no regular polyhedron whose faces are hexagons. (Hint: what does $\frac{1}{p} + \frac{1}{q} > \frac{1}{2}$ demand when $p = 6$?)

REFLECTION *No answer key. Write a few sentences.*

1. Pick the problem above that fought you hardest. What *feature* of it should have whispered "double counting"? Write the trigger down in your own words, so future-you recognizes it on sight.

NUDGES

Launch. (1) Each digit hits each place $2! = 2$ times: $2 \times (1+2+3) \times 111 = 2 \times 6 \times 111 = 1332$. (2) $\sum \deg = 2E \Rightarrow 7 \cdot 4 = 28 = 2E$, so $E = 14$; all-degree-3 is impossible since $7 \cdot 3 = 21$ is odd but $2E$ is even.

Core. (3) K_n has $\binom{n}{2}$ edges; counting by smaller endpoint gives $1 + \cdots + (n-1)$. (4) Largest element $i+1$ leaves $\binom{i}{r}$ ways below it. (5) Committee, captain, co-captain: pick the 2 leaders first ($\binom{n}{2}$ ways) then any subset of the other $n-2$ (2^{n-2}), so $\sum_k \binom{n}{k}\binom{k}{2} = \binom{n}{2}2^{n-2}$; the same two-role staging gives $\sum_k k^2\binom{n}{k} = n(n+1)2^{n-2}$.

Stretch. (6) Edges through faces: $2E = 5p + 6h$. Three faces (hence three edges) per vertex: $3V = 2E$. With $F = p + h = 12$, Euler $V - E + F = 2$ forces $5p + 6h = 60$, and with $p + h = 12$ this gives $p = 12$ (so $h = 0$): the solid must be a dodecahedron. (7) Split the committee by how many seniors it uses.

The Flash. (8) Total wins = number of games = $\binom{n}{2}$ (each game yields exactly one win). (9) $6 \times 9 = 54$ visible faces (six faces, nine unit faces each).

Proof Lab. (10) Double count pairs of clubs. Each unordered pair of clubs shares exactly one member, so the number of such pairs is $\binom{c}{2}$; grouping the pairs instead by their shared member, a person in d_p clubs accounts for the $\binom{d_p}{2}$ club-pairs meeting at them, and every pair is counted exactly once, giving $\sum_p \binom{d_p}{2} = \binom{c}{2}$. It is only a *first* step toward Fisher because it ties the club-degrees to c but says nothing yet about the number of people; turning it into a clubs-versus-students bound needs an extra regularity or Cauchy–Schwarz input. (11) A regular solid with p-gon faces and q meeting at each vertex needs $\frac{1}{p} + \frac{1}{q} > \frac{1}{2}$ (the angles around a vertex must total less than 2π). Setting $p = 6$ forces $\frac{1}{q} > \frac{1}{2} - \frac{1}{6} = \frac{1}{3}$, so $q < 3$ — impossible, since at least 3 faces must meet at any vertex. Hexagons can only tile the flat plane (the honeycomb), never close up into a solid.

CHAPTER TWO

2

Too Many Pigeons

If you have more pigeons than holes, some hole is hosting a party. The most obvious idea in mathematics — and one of the most devastating.

Pigeonhole: too many objects, too few labels.

You can know the following without taking attendance. It sounds impossible until you see why.

WONDER THIS

At any party — any party at all, anywhere, with at least two people — there are *always* two guests who have shaken hands with exactly the same number of other guests present.

You don't get to ask who anyone knows. You don't get to see the room. **How can this possibly be guaranteed?**

CURIOUS CARL • FIRST GUESS

Carl: "Two of thirteen must share a birth month? I'll test the pairs one at a time until a match turns up — how long can that possibly take?"

The catch: no searching required, Carl — 13 people dropped into only 12 months *forces* a repeat before you've compared a single pair.

It feels like it shouldn't be knowable. Maybe everyone shook a different number of hands? Let's chase that possibility and watch it collapse.

Say there are n people at the party. The number of hands a single guest could have shaken ranges from 0 (a wallflower who greeted no one) up to $n-1$ (a social butterfly who greeted everybody else). That's n possible "handshake counts" (the values $0, 1, 2, \ldots, n-1$) spread among n people. So far, no contradiction: n values, n people, everyone could be different.

But look closer at the two extremes. If somebody shook hands with *everybody* (count $n-1$), then nobody can have shaken hands with *nobody* (count 0): that social butterfly shook *their* hand too. The values 0 and $n-1$ cannot both appear. So the handshake counts really live in a set of only $n-1$ possible values, not n. And now we have n people trying to fit into $n-1$ slots. Two of them must land in the same slot. **Two guests shook the same number of hands.** Guaranteed, every time, no attendance required.

THE TURN *from grind to insight*

The head-on approach to a guarantee like this is to *search* for the coinciding pair — imagine every possible party, every assignment of handshake counts to guests, and check that you can never sneak all n people onto distinct numbers. With even a modest crowd the cases explode past any hope of listing, and "I couldn't build a counterexample" is not a proof anyway. **The turn** is to stop hunting altogether and simply *count two things*: n guests, but only $n-1$ available handshake values. The moment the items outnumber the boxes, two guests are *forced* to share one — existence proved with zero search. The lesson: never construct the coincidence — corner it by counting.

Stated plainly, the principle is almost embarrassingly simple:

THE MOVE

If you place more objects into boxes than there are boxes, then *some box holds at least two objects.* More sharply: if n objects go into k boxes, some box holds at least $\lceil n/k \rceil$ of them. That's it. The whole art is in *deciding what the pigeons are and what the holes are* — because the principle itself never does anything surprising; your *choice of boxes* does.

The principle is trivial. Choosing the boxes is not. A good **pigeonhole** proof feels like a magic trick because the boxes seem to appear from nowhere. That is the skill.

A pigeonhole proof is a compression proof. You take many objects and *compress* them into too few labels; the collision is then forced, for free. So the whole art is *designing the compression*, and good boxes beat bad boxes by a mile. The contrast is worth memorizing: for points in a unit square, splitting into two halves is too coarse (two points in one half can still be far apart), while four sub-squares are exactly right (two in one sub-square are guaranteed close). For a divisibility problem, the numbers themselves make hopelessly many labels, but their *remainders* mod m make exactly m. **Bad boxes prove nothing; the right compression proves everything.** You'll watch both of these payoffs land in a moment.

Drawing the right boxes

The friends-at-a-party proof already shows the move: the boxes were "possible handshake counts," and the only cleverness was noticing that two of them (0 and $n - 1$) exclude each other, squeezing n pigeons into $n - 1$ holes. Let's see the same move in a setting that looks geometric but is really pigeonhole in disguise.

WATCH THIS — **#1 Five points in a square**

Claim. Drop any 5 points inside a unit square. Two of them are guaranteed to lie within distance $\frac{\sqrt{2}}{2}$ of each other.

The boxes. Slice the unit square into four smaller squares, each $\frac{1}{2} \times \frac{1}{2}$.

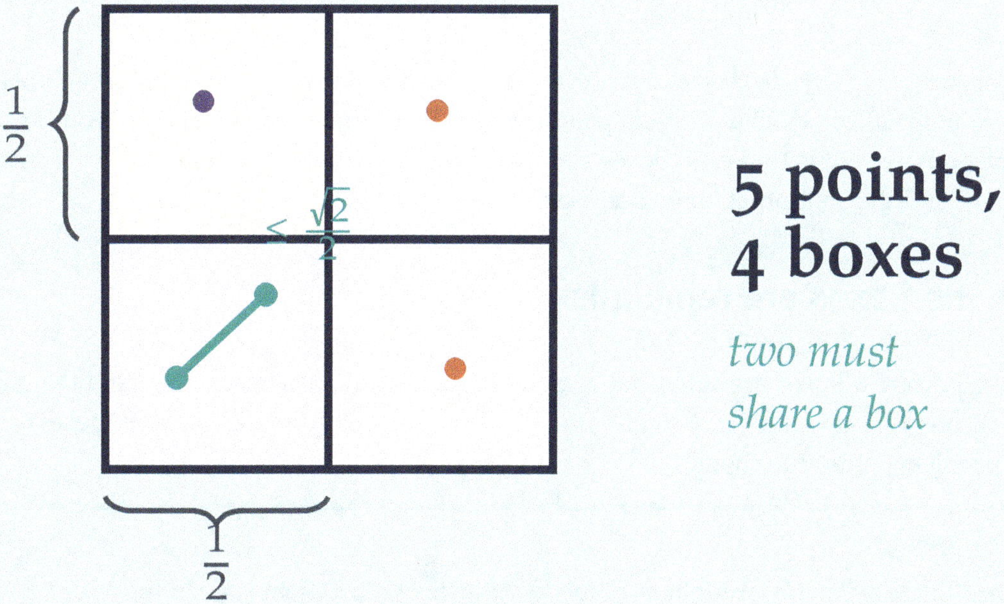

The pigeons. The 5 points. With 5 points and 4 boxes, some box swallows at least two of them.

The finish. Two points inside the same $\frac{1}{2} \times \frac{1}{2}$ square can be at most a diagonal apart, and that diagonal measures $\sqrt{(\frac{1}{2})^2 + (\frac{1}{2})^2} = \frac{\sqrt{2}}{2}$. So those two points are within $\frac{\sqrt{2}}{2}$. Done. ■

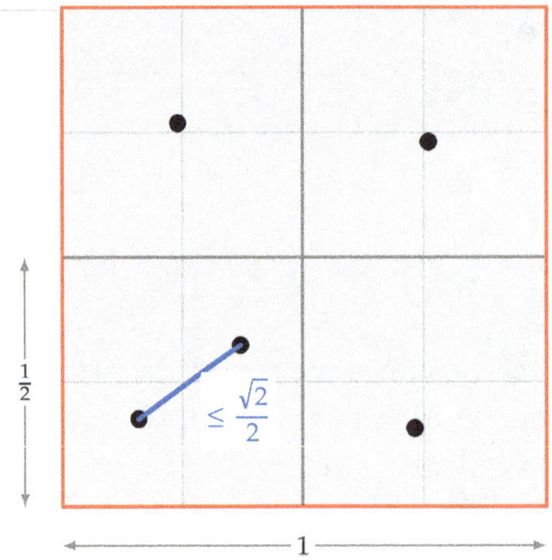

Five points in a unit square: two must share one $\frac{1}{2} \times \frac{1}{2}$ cell, so their distance is at most $\frac{\sqrt{2}}{2}$.

Trigger A claim that "two of these must be close" or "two must share a property," with more objects than natural categories.

Do first Carve the space (or the values) into exactly one fewer box than you have objects.

Trapdoor Boxes that are too big to force the conclusion, or so many boxes that pigeonhole says nothing. Aim for $k = n - 1$ boxes.

Upgrade Replace "4 sub-squares" with "m^2 sub-squares" to push two of $m^2 + 1$ points within distance $\frac{\sqrt{2}}{m}$.

The whole proof was the *decision* to cut the square into four. Nothing else required an idea. This is the rhythm to internalize: when a problem promises that "two things must collide," your job is not to do clever computation. It is to design boxes so that there is one fewer box than there are things. Get the boxes right and the contradiction writes itself.

When the boxes are remainders

The most powerful boxes are often not regions of space but *remainders*. Divide by m and there are exactly m possible remainders: $0, 1, 2, \dots, m-1$. That is a ready-made set of m boxes, and number-theory problems pour into them.

WATCH THIS **#2 A hidden multiple**

Claim. Take any 5 whole numbers. Some *consecutive block* of them (one of them, or two in a row, or three in a row, ...) adds up to a multiple of 5.

The trick: prefix sums. Write your numbers as a_1, a_2, a_3, a_4, a_5. Build the running totals

$$S_0 = 0, \quad S_1 = a_1, \quad S_2 = a_1 + a_2, \quad \dots, \quad S_5 = a_1 + \cdots + a_5.$$

That's 6 running totals: $S_0, S_1, \dots, S_5$.

The boxes. Their remainders modulo 5. There are only 5 possible remainders, but we have 6 running totals.

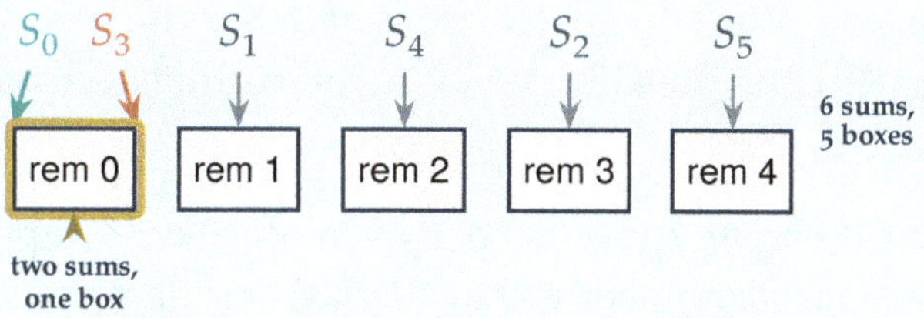

The finish. By pigeonhole, two of the running totals — say S_i and S_j with $i < j$ — land in the same remainder box. Their difference $S_j - S_i = a_{i+1} + a_{i+2} + \cdots + a_j$ is then divisible by 5. And that difference is exactly a consecutive block of the original numbers. ■

The *prefix-sum-plus-remainders* combination is one of the most reusable weapons in all of contest mathematics: it converts "some consecutive chunk has a divisibility property" into a one-line pigeonhole. Nothing about the five numbers mattered. They could be enormous, negative, repeated. Six sums, five boxes, done. Notice the recurring shape of these arguments: we never compute the block that works. Pigeonhole is a pure *existence* engine. It swears two pigeons share a hole; it serenely refuses to tell you which two. That is both its power and, occasionally, its frustration.

△ TRAPDOOR

Pigeonhole proves something *exists*; it never hands you the thing. If a problem asks you to *find* the colliding pair or to *construct* an example, pigeonhole alone won't finish it — you'll need an actual construction. Don't mistake "two must share a box" for "here are the two." Knowing the difference saves you from confidently writing a proof that answers a question nobody asked.

CURIOUS CARL • TRAPDOOR!

For 6 running totals modulo 5, Carl sets up his boxes: "Remainders $1, 2, 3, 4, 5$ — five boxes, six sums, so two must collide!" He starts sorting.

A pigeon flutters over and inspects the mailboxes: *"You forgot remainder* 0*, and you invented a remainder* 5 *that never happens — your boxes don't even fit the sums."* True: the remainders are $0, 1, 2, 3, 4$. Carl's count came out right by luck, but always list your boxes exactly, or an off-by-one hole will bite you.

The showpiece: order hiding inside chaos

Now for something that feels genuinely magical: pigeonhole can force *structure* to appear in a sequence that looks utterly random.

WONDER THIS

Line up any 10 distinct numbers in any order you like — shuffle them, scramble them, be as adversarial as you wish.

I claim that hidden inside your sequence there is *always* a run of 4 numbers (not necessarily next to each other) that is either steadily increasing or steadily decreasing. **You cannot avoid it.**

This is the **Erdős–Szekeres theorem**, and it is one of the most beautiful applications of pigeonhole ever found. The reason 10 is the magic number is that $10 = 3^2 + 1$: once you exceed 3×3, order becomes unavoidable.

WATCH THIS #3 Monotone runs are inevitable

Claim. Any sequence of $mn + 1$ distinct numbers contains an increasing run of length $m + 1$ or a decreasing run of length $n + 1$. (Our opening puzzle is $m = n = 3$, giving $3 \cdot 3 + 1 = 10$ and runs of length 4.)

The labels. To each position i in the sequence, attach a *pair of numbers* (a_i, b_i):

- a_i = the length of the longest *increasing* run *ending* at position i,
- b_i = the length of the longest *decreasing* run *ending* at position i.

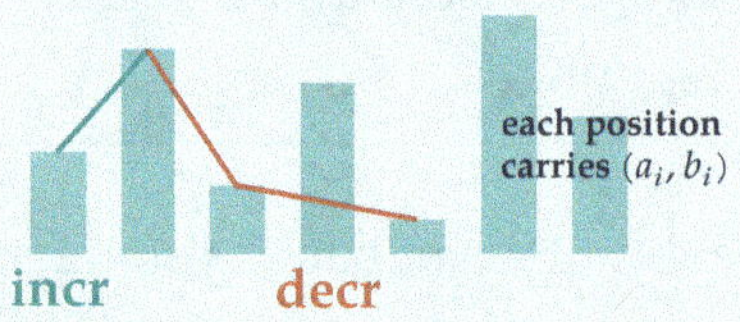

The boxes. Suppose, for contradiction, that there is *no* increasing run of length $m + 1$ and *no* decreasing run of length $n + 1$. Then every a_i lies in $\{1, \dots, m\}$ and every b_i lies in $\{1, \dots, n\}$. So each pair (a_i, b_i) has at most $m \times n$ possible values — that's our mn boxes.

The pigeons. The $mn + 1$ positions. More positions than boxes, so two positions $i < j$ share the same pair: $(a_i, b_i) = (a_j, b_j)$.

The contradiction. But two positions can *never* share a pair! Since the numbers are distinct, either $x_i < x_j$ or $x_i > x_j$. If $x_i < x_j$, then any increasing run ending at i extends to j, forcing $a_j > a_i$. If $x_i > x_j$, then $b_j > b_i$. Either way the pairs differ — contradicting that they were equal. So our assumption was false: a long monotone run must exist. ∎

Stop and admire what just happened. We took a sequence with *no assumed structure whatsoever* and proved that structure must be lurking inside it, not by examining the sequence, but by labeling each position with two numbers and counting the labels. The pigeons were positions; the holes were pairs of run-lengths; and the moment there were more positions than possible pairs, order had nowhere to hide. This is pigeonhole at its most astonishing: a guarantee of *pattern* purchased with nothing but counting.

SEEN BEFORE?

The "attach a clever label to each object, then count the labels" move is the secret engine behind many pigeonhole proofs — and it rhymes with the incidence-counting of **Chapter 1** (Double Counting). There, we counted one set two ways; here, we count labels against positions. Both are the same instinct: *find the right thing to count, and the problem solves itself.* You'll meet monotone-sequence ideas again in **Chapter 17** (Graph Theory).

TRY TO BREAK IT

The theorem forces a monotone run of length 4 once you line up 10 numbers. So the bound had better be tight — try to *break* it with only 9. Can you arrange nine distinct numbers with no increasing *and* no decreasing run of length 4? Push at it, then try the blocked staircase $3,2,1,\ 6,5,4,\ 9,8,7$: every increasing pick grabs at most one term per descending block, every decreasing pick stays trapped in a single block. It just barely dodges 4 — showing $10 = 3^2 + 1$ is exactly the threshold, not one more.

Payoff: two subsets, one sum

Let's close with a puzzle whose answer feels far too strong for how little we'll do to earn it.

WONDER THIS

Choose any 10 different numbers from 1 to 99.

I promise you can always split off *two* separate groups from your ten numbers — with no number shared between the groups — that have the *exact same sum*. Pick the ten numbers however you like; the matching pair of groups is always there.

The mechanism is pure pigeonhole, and the boxes are sums.

WATCH THIS **#4 Too many subsets, too few sums**

The pigeons. All the *subsets* of your 10 chosen numbers. A set of 10 numbers has $2^{10} = 1024$ subsets; ignore the empty one, leaving 1023 nonempty subsets.

The boxes. The *possible sums* of those subsets. The smallest possible sum is small, and the largest possible sum uses the ten biggest available numbers, $90 + 91 + \cdots + 99 = 945$. So every subset sum lands somewhere in $\{1, 2, \ldots, 945\}$ — at most 945 boxes.

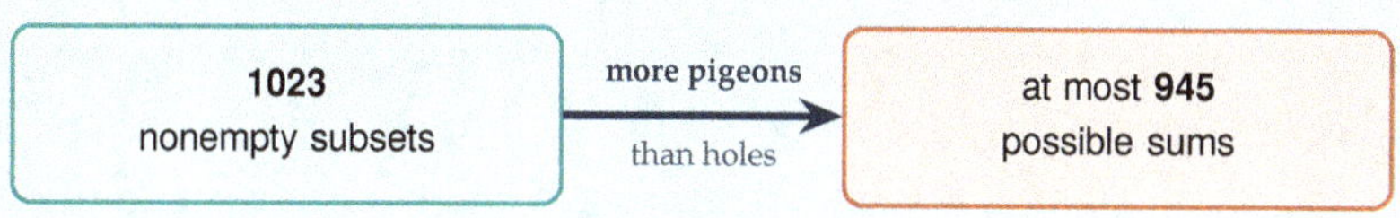

The finish. With 1023 subsets and only 945 possible sums, two *different* subsets A and B must share a sum. They might overlap — but just throw away the numbers they have in common. Removing the same numbers from both keeps their sums equal, and now the two leftover groups are disjoint and *still* have the same sum. ∎

There is no formula here, no construction, no clever algebra. Just the brute observation that 1023 pigeons cannot fit into 945 holes. The two equal-sum groups are out there in every choice of ten numbers, and pigeonhole knows it without ever finding them. *Once again: maddening, and magnificent.*

What to try when you're stuck

When a problem guarantees that "two things must coincide," "some thing must repeat," or "a pattern must appear," reach for pigeonhole and run this checklist:

1. **Identify the pigeons.** What are the objects there are many of?
2. **Invent the holes.** Remainders? Sub-regions? Pairs of labels? Possible sums? Aim for *one fewer hole than pigeon*, or close to it.
3. **Check the count.** Confirm pigeons > holes. If not, redesign the holes — merge them, or find two that secretly exclude each other.
4. **Read off the collision.** Two pigeons in one hole means two objects share a property. Translate that back into the problem's language.
5. **Remember the limit.** Pigeonhole proves existence, never location. If you need the actual pair, you need more.

THE MOVE

One-page recap. Pigeonhole: more objects than boxes ⇒ some box holds ≥ 2 (and in general $\geq \lceil n/k \rceil$). The principle is free; the *boxes* are the idea. Master toolbox: **exclude two extreme categories** (friends at a party), **sub-regions** (points in a square), **remainders mod m** (prefix sums), **pairs of labels** (Erdős–Szekeres), and **possible sums** (equal-sum subsets). When stuck, count the pigeons, count the holes, and make the second number smaller.

THE NAME FOR IT

The Pigeonhole Principle

If more objects are placed than there are boxes, some box holds at least two. Mathematicians call this the *Pigeonhole Principle* (or Dirichlet's box principle), and its sharpened form — n objects in k boxes force some box to hold at least $\lceil n/k \rceil$ — is the *generalized pigeonhole principle.*

WHERE IT LEADS The pigeonhole principle is also known as *Dirichlet's box principle*, after the analyst who used it to prove deep facts about how well irrationals can be approximated by fractions (*Dirichlet's approximation theorem*). Its infinite cousins underpin compactness arguments throughout real analysis and the theory of *equidistribution*.

GOING FURTHER *optional • beyond AIME*

This box is an optional detour — the result below reaches past contest level, and nothing later depends on it.

Dirichlet's approximation theorem. Every irrational number can be pinned down by fractions startlingly well. Precisely: for any irrational α and any whole number N, there is a fraction p/q with $1 \le q \le N$ such that

$$\left|\alpha - \frac{p}{q}\right| < \frac{1}{qN}.$$

Since $q \le N$, the bound $\frac{1}{qN}$ is at least as strong as $\frac{1}{q^2}$, so the fraction pins α down on the *scale of* $1/q^2$ — far sharper than rounding α to the nearest tenth or hundredth would give.
The proof is pure pigeonhole, and it reuses the prefix-sum-plus-remainders instinct from this chapter — only now the "boxes" are slices of the unit interval. Consider the $N+1$ fractional parts

$$\{0\cdot\alpha\},\ \{1\cdot\alpha\},\ \{2\cdot\alpha\},\ \ldots,\ \{N\cdot\alpha\},$$

each lying in $[0,1)$. Cut $[0,1)$ into N equal subintervals of length $1/N$. That is $N+1$ pigeons (the fractional parts) and only N holes (the subintervals), so two of the fractional parts, say $\{i\alpha\}$ and $\{j\alpha\}$ with $i < j$, must land in the same subinterval and therefore differ by less than $1/N$. Setting $q = j - i$ (so $1 \le q \le N$) and letting p be the difference of the two integer parts, this says $|q\alpha - p| < 1/N$; dividing by q gives $|\alpha - p/q| < 1/(qN)$. The whole theorem is one well-chosen family of pigeons dropped into N holes.

CURIOUS CARL • EUREKA

Carl: "I can prove two of us share a birthday and still have no clue *which* two. That should bother me more than it does... actually, I think I love it."

SAME BEAST, NEW COSTUME

The same "more objects than labels" beast, in costume across the book:

Same remainder → Erdős–Szekeres → Dirichlet approximation → Ramsey $R(3,3)$.

Each drops a crowd of pigeons into too few holes — remainders, run-length pairs, subintervals, colored triangles — and reads off the forced collision. The whole art is only ever choosing the holes.

The practice ladder

THE PRACTICE LADDER *climb at your own pace*

LAUNCH *Warm the engine. Nudges below.*

1. A drawer holds socks of 3 colors, jumbled in the dark. How many socks must you grab to be *sure* of a matching pair?

2. Show that among any 13 people, at least two share a birth *month*.

CORE BUILDER *Master the central moves.*

1. Among any $n + 1$ numbers chosen from $\{1, 2, \dots, 2n\}$, prove two of them are consecutive (and hence coprime).

2. Among any $n + 1$ numbers chosen from $\{1, 2, \dots, 2n\}$, prove one of them divides another. (Boxes: the *odd part* of each number.)

3. Prove that some multiple of 2025 has only 0s and 1s in its decimal digits. (Boxes: remainders of $1, 11, 111, \dots$ mod 2025.)

STRETCH *Transfer to unfamiliar ground.*

1. Six people are at a party. Prove that either three of them are mutual friends, or three are mutual strangers. (You'll meet this again in Chapter 17 as the first theorem of Ramsey theory — here, prove it with pigeonhole.)

2. Prove that any sequence of $n^2 + 1$ distinct reals has a monotone subsequence of length $n + 1$. Then show this bound is sharp: for $n = 3$, exhibit a sequence of length $n^2 = 9$ with *no* monotone subsequence (increasing or decreasing) of length $n + 1 = 4$.

THE FLASH *No clock — each cracks open in one flash of insight, once you find the angle. Nudges at the back.*

1. Seventeen points sit inside an equilateral triangle of side 1. Must two of them be within $\frac{1}{4}$ of each other?

2. Pick any 51 numbers from $\{1, 2, \dots, 100\}$. Must two of them sum to 101?

PROOF LAB *Write it clean, in full sentences.*

1. Prove that any 10 distinct two-digit numbers contain two disjoint subsets with equal sum.

REFLECTION *No nudge. Write a few sentences.*

1. Of the problems above, which had the least obvious choice of boxes? Describe, in your own words, the clue that told you what the holes should be.

NUDGES

Launch. (1) Four socks — three could be one of each color, the fourth forces a match. (2) 13 people, 12 months. **Core.** (3) Pair up $\{1,2\}, \{3,4\}, \dots$; n pairs, $n+1$ picks. (4) Write each number as $2^a \cdot (\text{odd})$; only n odd parts available. (5) Among $1, 11, 111, \dots$ two share a remainder mod 2025; subtract. **Stretch.** (6) Fix one person; of the other five, three are friends-or-strangers with them, then chase the triangle. (7) The blocked staircase $3, 2, 1, 6, 5, 4, 9, 8, 7$ (length $9 = 3^2$) works: any increasing subsequence takes at most one term per descending block ($3 \to 6 \to 9$), and any decreasing subsequence stays inside one block $(3, 2, 1)$, so the longest monotone subsequence has length exactly 3 — none reaches 4, proving $n^2 + 1$ is sharp. **The Flash.** (8) Cut the triangle into 16 small triangles of side $\frac{1}{4}$. (9) Pair $\{1, 100\}, \{2, 99\}, \dots$ into 50 boxes. **Proof.** (10) The 10 numbers have $2^{10} = 1024$ subsets, yet every subset sum lies between 0 and $90 + 91 + \cdots + 99 = 945$ — fewer than 1024 possible values — so by pigeonhole two distinct subsets A, B share the same sum. Deleting the common elements $A \cap B$ from each leaves two *disjoint* subsets that are still nonempty (since $A \neq B$) and still have equal sums, the required pair.

Chapter Three

3

What Can’t Be Undone

Some things never change no matter what you do to them. Find that one unchanging thing, and you can predict the impossible.

Invariants: watch what cannot change.

A chessboard, and a single lonely knight.

WONDER THIS

A knight sits on a corner square of a chessboard. It hops around in its usual L-shaped way — and after exactly 2025 moves, it tries to land back on the very square it started from.

Can it? No matter how cleverly it wanders, can a knight return home in exactly 2025 moves?

CURIOUS CARL • FIRST GUESS

Carl: “So I *follow* the knight — trace every hop, branch at every choice, and see whether some 2025-move tour ever finds its way home.”

The catch: billions of branches, Carl — brute force can’t even dent this. The win is to stop chasing the knight and watch one thing it *can’t* control.

You could try to trace knight paths for an hour and get nowhere: there are billions of possibilities. But there's a single observation that settles it instantly, and it has nothing to do with paths at all. It has to do with *color*.

Color the chessboard black and white in the usual checkerboard pattern. Now watch a single knight move: a knight always jumps from a square of one color to a square of the *other* color. Every. Single. Time. Two-over-and-one-up never lands you on the same color you left. So the knight's color *flips* with every move: white, black, white, black, like a metronome.

After an *even* number of moves, the knight is back on its starting color. After an *odd* number of moves, it's on the opposite color. To return home, the knight needs to be on its starting square (which is, of course, its starting color), so it needs an *even* number of moves. But 2025 is odd. **The knight cannot possibly return home in 2025 moves.** And we proved it without tracing a single path.

That color is an **invariant**: a quantity that the rules of the game can never change. Here the rules flip the color every move, so the color's *parity* (its even-or-oddness across the whole journey) is locked to the number of moves. Find the right invariant and entire oceans of possibilities drain away, leaving only what's actually achievable.

THE TURN *from grind to insight*

The obvious attack is to chase the knight: list its moves, branch on every choice, and hunt for a 2025-step path that comes home. The tree of possibilities has billions of branches, and you would never reach the bottom of it. **The leap** is to stop following the path *at all* and instead watch one quantity the moves cannot help changing — the square's *color*. A single remark about that one quantity decides a question brute force cannot even dent. When a process is too tangled to follow, don't follow it — find what it can't change.

THE MOVE

Two tools, one family. This chapter is built on a pair of close cousins, and it pays to keep them straight:

- An **invariant** is a quantity that *never changes* under the moves. Use it to prove a target is *unreachable*: if the start and the goal disagree about the invariant, no sequence of moves can ever connect them.
- A **monovariant** is a quantity that can only move in *one direction* (say, never increases). Use it to prove a process *must stop* — but it has to fall in *discrete* steps: a *non-negative integer* (or any quantity that drops by at least a fixed positive amount each move) can't fall forever. Merely "bounded below" is not enough — a real number can shrink endlessly, as in $1, \frac{1}{2}, \frac{1}{3}, \ldots$.

Same instinct — watch one well-chosen quantity — aimed at two different questions: *can we get there?* versus *must this end?* The whole art is *inventing the quantity to watch.*

Parity: the simplest invariant of all

The knight argument was powered by the humblest invariant in mathematics: *parity*, the even-or-odd of a number. Parity is a wonderful first thing to try because it is so cheap to check and so often decisive. If a process changes some count by 2 each step (or flips something an even number of times), then parity is frozen, and frozen quantities are exactly what prove impossibility.

WATCH THIS — #1 The mutilated chessboard

Claim. Snip off two *opposite* corners of an 8 × 8 chessboard, leaving 62 squares. You cannot tile those 62 squares with 31 dominoes, each covering two side-by-side squares.

The invariant: color balance. Color the board as a checkerboard. Every domino, wherever you place it, covers exactly one black square and one white square — because adjacent squares always differ in color.

The contradiction. The two opposite corners are the *same* color (check: corners on a diagonal always match). Removing them leaves 32 squares of one color and only 30 of the other. But 31 dominoes must cover 31 black and 31 white — a perfect balance. Since $32 \neq 30$, no tiling can exist. ■

32 dark · **30** light

Remove two opposite (same-colour) corners: 32 dark and 30 light squares remain, so no domino tiling can exist.

Trigger A tiling or covering question where the pieces always cover a fixed "color signature" (here, one of each color).

Do first Color the board and count how many of each color the pieces *must* cover, then compare with what's available.

Trapdoor Assuming a balanced region can always be tiled — color balance is *necessary*, not sufficient. It can prove "impossible" but never, by itself, "possible."

Upgrade Swap the 2-coloring for a 3- or 4-coloring tuned to a bigger tile; the same imbalance argument kills L-trominoes and long bars (see Chapter 1's coloring cousins).

Here's the rhythm that makes parity arguments feel effortless once you've internalized them: *find the quantity the move can't change, then show the start and the goal disagree about it.* The mutilated board can't be tiled, not because tiling is hard, but because the color count is frozen at an imbalance the dominoes can never produce. The same one-line idea handles a whole zoo of "erase and replace" games.

TRY TO BREAK IT

Color balance is *necessary* for a domino tiling — equal black and white. But is it *sufficient*? Hunt for a region with equal numbers of black and white squares that still *cannot* be tiled by dominoes. (*Hint: a bent or pinched shape can balance the colors yet strand a square with no available partner.*) When you find one, you'll know in your bones why "the invariant allows it" never means "it happens."

WATCH THIS — **#2 The vanishing blackboard**

Claim. The numbers $1, 2, 3, \ldots, 2024$ are written on a board. Repeatedly, erase any two of them, say a and b, and write back their difference $|a-b|$. After 2023 erasures, a single number remains. It is always *even*.

The invariant: the parity of the total sum. Erasing a, b and writing $|a-b|$ changes the running total by

$$(a+b) - |a-b| = 2\min(a,b),$$

which is always *even*. So the parity of the sum of everything on the board never changes, from the first move to the last.

The finish. The starting sum is $1+2+\cdots+2024 = \frac{2024\cdot 2025}{2} = 2{,}049{,}300$, which is even. The lone survivor equals the final sum, whose parity is still even. So the last number standing must be *even* — guaranteed, no matter which pairs you erased or in what order. ■

Notice the move never told us *which* number survives, just its parity. That's the recurring flavor of invariant arguments: they pin down a property of the outcome while staying gloriously indifferent to the messy details of how you got there.

△ TRAPDOOR

An invariant can only prove a target is *unreachable* (start and goal disagree) — it can never, on its own, prove something *is* reachable. To show a state *can* be achieved, you must actually exhibit a sequence of moves that gets there. Confusing "the invariant allows it" with "it happens" is the most common invariant blunder. Allowed is not the same as achieved.

CURIOUS CARL • TRAPDOOR!

Carl sets up a move-puzzle where the start and the goal *agree* on the parity invariant. "They match — so I can *definitely* reach the goal!" he declares, and starts making moves.

The goal state folds its arms: *"Agreeing on the invariant only means you're not forbidden. It doesn't hand you a route — you still have to build one."* Carl nods slowly. An invariant can slam a door shut; it never swings one open for you.

Choosing a smarter invariant

Plain even-versus-odd is just the case "modulo 2." When a move shifts things by a fixed amount, the right invariant often lives at a *different* modulus — and choosing it well is where the craft comes in. The trick: pick the modulus that makes the move act *uniformly*, so something is forced to stay constant.

WATCH THIS — #3 Three piles that can't both empty

Claim. Three piles hold 11, 12, and 13 chips. A move removes one chip from *two* of the piles and adds *two* chips to the third. Starting from $(11, 12, 13)$, you can never reach a state where two piles are simultaneously empty.

The invariant: residues mod 3. A move sends a pile triple (x, y, z) to something like $(x - 1, y - 1, z + 2)$. Look at each coordinate modulo 3: it changes by $-1, -1$, and $+2 \equiv -1 \pmod 3$. So *every* pile's residue drops by exactly 1 mod 3 — a uniform shift.

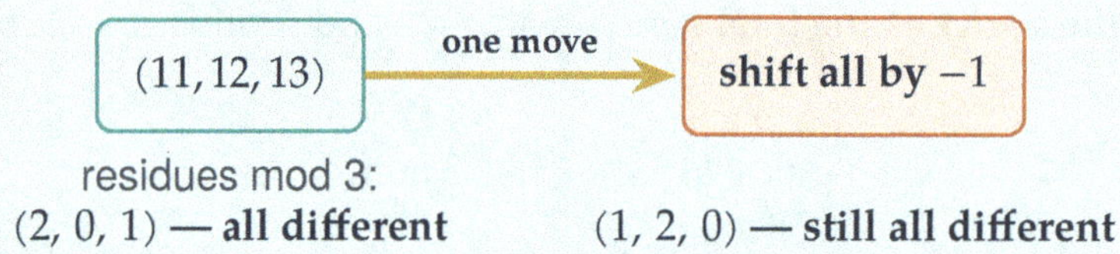

The contradiction. Because all three residues shift together, any two residues that start *different* stay different forever. At the start, $(11, 12, 13)$ has residues $(2, 0, 1)$ — all distinct. But "two piles empty" means a state like $(N, 0, 0)$, whose residues $(N \bmod 3,\ 0,\ 0)$ include *two equal* values. Distinct residues can never become equal under a uniform shift. So two piles can never both be empty. ■

The art was choosing modulus 3: the smallest one that makes the move "$-1, -1, +2$" collapse into the clean uniform shift "$-1, -1, -1$." Pick the modulus that simplifies the move, and the invariant reveals itself. Try modulus 2 here and you'd see nothing useful; the move's arithmetic is secretly a multiple-of-3 story, so 3 is the modulus that listens.

Coloring: an invariant you paint onto the board

Here is the same idea wearing its most visual costume. When a problem is about *tiling* (covering a board with pieces of a fixed shape), the winning move is to *color* the cells so cleverly that every piece, wherever it lands, always covers the same balanced palette. Then a simple color count becomes an unbreakable invariant, and a mismatch proves the tiling impossible.

WONDER THIS

Take a 5×5 board and a supply of straight 1×3 tiles (each covers three cells in a row or column).

Can you tile the whole board? Twenty-five cells, tiles of three — already $3 \nmid 25$, so *exactly* tiling is hopeless. But sharpen the question: cover 24 cells and leave one empty. **Which single cell could be the leftover?**

Counting alone ($25 = 8 \times 3 + 1$) tells you one cell must be left uncovered, but not *which*. For that we need a coloring tuned to the tile.

WATCH THIS

#4 Three diagonals for a straight tromino

The coloring. Color cell (x, y) with the value $(x + y) \bmod 3$ — three colors running in diagonal stripes. The magic: a straight 1×3 tile, placed horizontally or vertically, always covers three *consecutive* values of $x + y$, hence exactly one cell of each color $0, 1, 2$.

1	2	0	1	2
0	1	2	0	1
2	0	1	2	0
1	2	0	1	2
0	1	2	0	1

one tile: colors 0, 1, 2

The count. Across the 25 cells, color 0 appears 9 times while colors 1 and 2 appear 8 times each. Every tile removes one of each color, so after placing eight tiles you have removed eight of each — leaving 1 cell of color 0 and 0 of colors 1, 2. Therefore *the single uncovered cell must have color* 0: $(x + y) \equiv 0 \pmod 3$. The coloring didn't just say "impossible"; it *narrowed* the hole to those nine cells — though that's only a *necessary* condition. (A second coloring, in the practice ladder, tightens it all the way down to the center.) ■

The creative leap is choosing the modulus to *match the tile*. A checkerboard (two colors) would have told us nothing here: a straight tromino lands unevenly on it depending on position. The tile has length 3, so the coloring must speak modulo 3. This is the same "pick the right modulus" instinct as the three-piles problem, now painted onto a grid.

△ TRAPDOOR

The classic warm-up: remove two *opposite* corners of an 8×8 board and try to tile the rest with 1×2 dominoes. It looks plausible — 62 cells, 31 dominoes — but a checkerboard coloring kills it instantly: every domino covers one black and one white cell, yet opposite corners share a color, so you've removed two of the *same* color and left 32 of one and 30 of the other. No tiling can fix an imbalance the pieces are blind to. Always check the color counts *before* hunting for a tiling.

When two colors are too coarse, escalate the palette. For a straight 1×4 tile, color the columns $x \bmod 4$ in four vertical stripes: a horizontal tile then covers one of each color, but a *vertical* tile covers four cells of the *same* color — and comparing the four color totals on a defective board exposes obstructions a two-coloring cannot see. The rule of thumb: start with two colors, and if the tile is balanced under them, raise the modulus, switch to stripes, or even attach real-number *weights* to cells until every tile sums to the same value. The right weighting turns "impossible" into a one-line inequality.

The showpiece coloring: rhombi on a triangular grid

The most beautiful coloring of all is not on a square grid at all. Take an equilateral triangle of side n, and slice it into n^2 small unit triangles by lines parallel to the three sides. Some of these little triangles point *up* ($\triangle$) and some point *down* ($\triangledown$). Now try to tile the whole figure with *rhombi*, where each rhombus is made of one up-triangle glued to an adjacent down-triangle.

WONDER THIS

For which side-lengths n can the triangular grid be completely tiled by these rhombi?

Try $n = 2$: four small triangles, and ... you cannot do it. **Why not — and what rules out every** n**?**

WATCH THIS **#5 Up minus down: the unbeatable count**

The two-coloring. Color every *upward* unit triangle one color and every *downward* one another. The key fact about the rhombus: each rhombus is exactly one up-triangle joined to one down-triangle, so *every* rhombus, no matter where it sits or how it is rotated, covers exactly one up and one down.

The count that ends it. Tally the unit triangles row by row (top row down). Reading off the figure, a side-n triangle contains $1 + 2 + \cdots + n = \frac{n(n+1)}{2}$ upward triangles but only $0 + 1 + \cdots + (n-1) = \frac{n(n-1)}{2}$ downward ones. The difference is

$$(\text{up}) - (\text{down}) = \frac{n(n+1)}{2} - \frac{n(n-1)}{2} = n.$$

There are always exactly *n more* upward triangles than downward. But every rhombus consumes one up and one down in lockstep, so any rhombus tiling would need the two counts to be *equal*. Since they differ by $n \geq 1$, **the triangular grid can never be tiled by rhombi — for any** n. ■

The proof is nothing more than that: paint up against down, and the figure betrays a surplus of n that no rhombus can ever absorb. No case analysis, no clever construction. Just one coloring and one subtraction. It is the purest illustration of the chapter's creed: *find the quantity the moves can't change, and the impossible announces itself.*

The showpiece: a quantity that can only fall

So far our invariants have been frozen. But there's a close cousin — the *monovariant* — that is just as powerful, and it answers a completely different question: not "is this reachable?" but **"must this process ever stop?"** A monovariant is a quantity that can only move in *one* direction. If it is a non-negative integer that strictly decreases — or drops by at least a fixed positive amount while staying bounded below — the process has nowhere left to go.

WONDER THIS

You're sorting a shuffled row of numbered cards. The only move you're allowed: find two *adjacent* cards that are out of order (a bigger one sitting just left of a smaller one) and swap them.

You keep swapping out-of-order neighbors. **Is it possible to get stuck swapping forever?** Or must this always grind to a halt?

It *feels* like it should stop, since the row looks more sorted each time. But "feels like" is not a proof, and there are processes that look like they're improving yet never terminate. We need a quantity that provably can only go one way.

WATCH THIS #6 Inversions: the disorder meter

The monovariant. Define an *inversion* to be any pair of positions where a larger number sits to the left of a smaller one (the two needn't be adjacent). Let I be the total number of inversions in the row — a single number measuring "how scrambled" the row is.

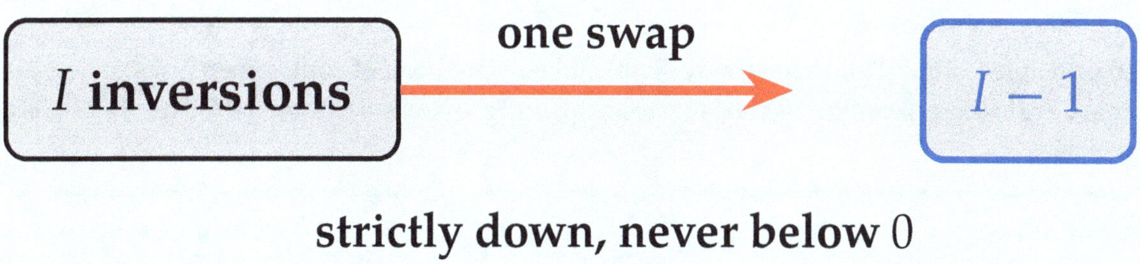

Why it can only fall. When you swap an adjacent out-of-order pair, you fix exactly that one inverted pair and disturb no others — every *other* pair's relative order is untouched. So each legal swap drops I by *exactly one*.

The finish. I is a whole number, it starts at some finite value, and it strictly decreases with every move. A non-negative integer cannot decrease forever — it would eventually go negative, which is absurd. So the process *must* stop, and it stops precisely when $I = 0$, that is, when the row is fully sorted. As a bonus, the largest possible number of swaps is the largest possible I, which for n cards is $\binom{n}{2}$ (every pair inverted, the totally-reversed row). ∎

This is the signature of a monovariant proof, and it's worth savoring: we proved that a process *must terminate* without knowing anything about which swaps happen or in what order. The disorder meter I can only tick downward, and downward has a floor. Whenever you're asked "does this always end?"

— a game, a rearrangement, a chip-firing, a sequence of operations — hunt for a non-negative integer that strictly drops each move. Find it, and termination is automatic.

SEEN BEFORE?

A monovariant is really an *extremal* idea in motion — "the quantity can't go below its minimum" is the same instinct as "take the smallest counterexample," which powers the induction and strong-induction arguments of **Chapter 5** and the descent proofs of number theory in **Chapter 9**. And the modular invariants here are the exact residue-counting from **Chapter 2**'s mod-m boxes, now watching a process instead of a static set.

The extremal principle: start at the edge

The monovariant worked by watching a quantity sink to its *minimum*. That points at a second great structural idea, a true peer of parity and invariants: when a problem feels infinite or shapeless, *look at the most extreme object* (the largest, the smallest, the closest, the leftmost) and let its extremeness do the work.

THE MOVE

The Extremal Principle. In any finite collection, *something is biggest and something is smallest.* Choose that extreme object and ask what its extremeness *forbids*. The minimum can have nothing below it; the maximum, nothing above; the closest pair, nothing closer. That "nothing" is often exactly the contradiction or the structure you need — reached without examining the other cases at all.

The principle is almost embarrassingly simple, like pigeonhole: of course there's a smallest element. The art, again like pigeonhole, is choosing *which* quantity to extremize so that the extreme case is forced to behave.

WATCH THIS — **#7 The party with no crossings**

The problem. On a field stand n red posts and n blue posts, no three in a line. Prove you can always run a rope from each red post to a distinct blue post so that *no two ropes cross*.

The extremal choice. There are finitely many ways to pair the reds with the blues (exactly $n!$ of them), so among all pairings there is one whose *total rope length is smallest*. Pick that one. We claim it has no crossings.

Let the minimum forbid a crossing. Suppose two ropes did cross — say red R_1 to blue B_1 and red R_2 to blue B_2, meeting at a point X. Re-pair them the other way: R_1 to B_2 and R_2 to B_1. By the triangle inequality (going straight beats going via X),

$$R_1B_2 + R_2B_1 < (R_1X + XB_2) + (R_2X + XB_1) = R_1B_1 + R_2B_2.$$

The swap *strictly shortens* the total length — contradicting that we chose the shortest pairing! So no two ropes cross. **The minimum-length pairing is automatically crossing-free.**

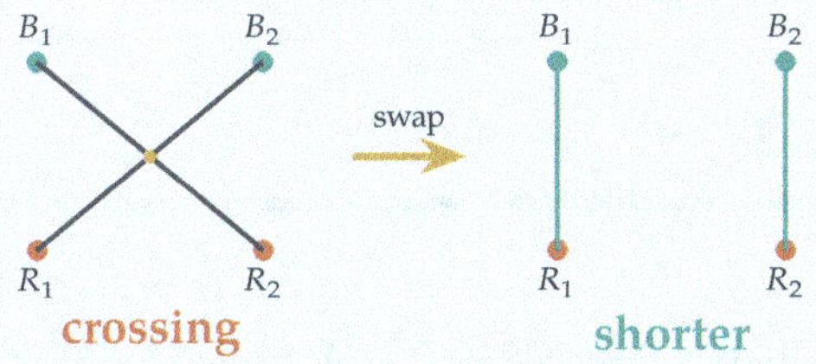

■

Trigger A problem about a configuration that must exist or have a property, especially with words like *always*, *some*, *no two*, or when the objects are finite but numerous.

Do first Pick the extreme object — the largest, smallest, closest, or longest — and ask what its extremeness rules out. Often a single forbidden move finishes the proof.

Trapdoor Choosing an extreme of the *wrong* quantity, so the extreme case isn't actually constrained. If the minimum doesn't forbid anything useful, extremize a different measurement.

Upgrade Extremal + "swap to improve" is the engine behind countless olympiad proofs; it's also the soul of the monovariant (the process halts *at* the extreme).

Notice the kinship with the monovariant we just met: there, a quantity *fell* to its minimum; here, we *start* at the minimum and read off what it forbids. **An extreme object is a monovariant that has already finished falling.** *When a problem sprawls, don't survey the whole landscape — walk straight to its edge and see what the edge makes impossible.*

GOING FURTHER *optional • beyond AIME*

This box is an optional detour, one notch past contest level; nothing later relies on it. The same "take the smallest distance" move that uncrossed the ropes settles a famous and far deeper-sounding question.

The Sylvester–Gallai theorem. Place finitely many points in the plane, not all on a single line. Then *some* straight line passes through *exactly two* of them. (Such a line is called an *ordinary line.*) The surprise is that you cannot arrange the points — however cleverly — so that every line through two of them is forced to scoop up a third.

The proof is extremal, and it is the rope-crossing idea wearing a heavier coat. Among *all* pairs (P, ℓ) where ℓ is a line through at least two of the points and P is one of the points *not* on ℓ, pick the pair whose distance from P to ℓ is smallest and positive — such a pair exists because the points are not all collinear, and there are only finitely many to compare. Now suppose, toward a contradiction, that this minimal line ℓ ran through three of the points. Drop the perpendicular from P to ℓ, meeting it at a foot F. Of the three points on ℓ, at least two lie on the same side of F (or one sits at F); call the nearer one X and a farther one Y. Comparing the two right triangles involved shows the distance from X to the line PY is *strictly smaller* than the distance from P to ℓ — a closer (point, line) pair, contradicting our choice of the minimum. So ℓ cannot hold a third point: it passes through exactly two. The whole theorem falls out of choosing the right thing to minimize, exactly as the shortest pairing could not cross.

Payoff: the coin-flipping puzzle

Back to a quick one, now that you have the full toolkit.

WONDER THIS

Twenty coins lie on a table, all showing heads. Your move: flip *any two* coins at once (your choice which two), as many times as you like.

Can you ever reach all twenty showing tails? And what if there were twenty-*one* coins?

Watch the number of heads. Flipping two coins changes the head-count by +2 (two tails became heads), by −2 (two heads became tails), or by 0 (one of each). In every case, the change is *even*. So the *parity* of the number of heads is an invariant.

With 20 coins we start at 20 heads (even) and want 0 heads (even). The parities agree, so the invariant raises no objection — and indeed it's easy to do: flip the coins two at a time, ten flips, all tails. But with 21 coins, all heads means we start at 21 heads (*odd*) and want 0 heads (*even*). The parity of the head-count can never change, so odd can never become even. **With an odd number of coins, all-tails is forever out of reach.** The humble parity invariant, one last time, draws the exact line between possible and impossible.

What to try when you're stuck

When a problem involves a *process* — moves, operations, a game, repeated replacement — and asks whether something is reachable or whether it must end, reach for invariants and monovariants:

1. **Write down exactly what one move does** to the configuration.
2. **Hunt for an invariant.** What quantity does the move leave *unchanged*? Try: a sum, a product, a parity, a coloring, residues mod m.
3. **For "unreachable" questions,** compare the invariant at the start and at the target. If they disagree, you're done.
4. **For "must it stop" questions,** hunt for a monovariant: a non-negative integer that strictly drops every move.
5. **For "it's possible" claims,** the invariant is not enough — actually build a sequence of moves that gets there.
6. **When a configuration must exist or behave,** forget the process and apply the extremal principle: pick the largest, smallest, or closest object and ask what its extremeness forbids.

THE MOVE

One-page recap. An invariant never changes; a monovariant moves one way; an *extreme* object is where a monovariant has stopped. Invariants prove *impossibility* (start and goal disagree);

monovariants prove *termination* (a bounded integer can't fall forever); the *extremal principle* proves *existence or structure* (the minimum forbids what lies below it). Toolbox: **parity** (knight color, head-count), **color balance** (mutilated board), **sum or product mod m** (vanishing blackboard, three piles), **inversion count** (sorting must stop), and **take the extreme** (the shortest pairing can't cross). When stuck, ask what the move *can't* change, what it can only push one way, and what the most extreme object makes impossible.

THE NAME FOR IT

Invariants, parity, and coloring arguments

A quantity left unchanged by every allowed move is an *invariant*; when that quantity is simply "odd or even," it is a *parity* argument. Painting a board so that every tile covers a fixed palette is a *coloring argument* — an invariant made visual, the standard route to *tiling-impossibility* proofs. A quantity that can only move one way (never increasing, say) is a *monovariant*, the standard tool for proving a process must terminate.

WHERE IT LEADS Invariants are the elementary face of *conserved quantities* — the same idea that, via *Noether's theorem*, ties every symmetry of a physical system to a conservation law (energy, momentum, charge). Coloring and parity arguments grow into $\mathbb{Z}/m\mathbb{Z}$ *homomorphisms* and the *coloring invariants* of topology and knot theory.

CURIOUS CARL • EUREKA

Carl sweeps his flashlight across a room of spinning gears and lands on the one dial that never moved. "Everything else is just noise trying to distract me. *You're* the one telling me it can't be done."

SAME BEAST, NEW COSTUME

The same "watch a conserved quantity" beast, in disguise across the book:

Parity of a head-count → Color balance of a tiling → Residue mod m of a pile → Inversion count of a shuffle → Permutation sign in the 15-puzzle.

Each is one number the rules leave alone (an invariant) or push only one way (a monovariant). Find that number and the impossible announces itself.

The practice ladder

THE PRACTICE LADDER

climb at your own pace

LAUNCH *Warm the engine. Nudges below.*

1. A frog jumps along a number line, each jump exactly 1 unit left or right. Can it return to its start after exactly 7 jumps?
2. A light switch starts OFF. Each move toggles it. After 100 moves, is it ON or OFF?

CORE BUILDER *Master the central moves.*

1. The numbers 1 through 7 are on a board. You may replace any two, a and b, with $a+b-1$. After six steps one number remains. What is it, and why is it forced?
2. On a 5 × 5 board, can a single rook visit every square exactly once and return to start, moving one square at a time? (Color argument.)
3. A jar holds 7 black and 8 white stones. Repeatedly remove two stones: if they are the *same* color, drop a *black* stone into the jar; if they are *different* colors, drop a *white* stone in. Each round removes one stone, so after 14 rounds a single stone remains. What color is it? Find the invariant that forces the answer.

STRETCH *Transfer to unfamiliar ground.*

1. In the 15-puzzle, the tiles 1 through 15 sit in order except that 14 and 15 are swapped. Prove this position is unreachable from the solved board. (Invariant: permutation parity plus the blank's position.)
2. Each cell of a 3 × 3 grid holds +1 or −1. A move flips all signs in one row or one column. Prove you cannot turn a single −1 (rest +1) into all +1.
3. (*A monovariant, not a sort.*) Among n people, some pairs are enemies. Prove you can split everyone between two rooms so that each person has *at most as many* enemies in their own room as in the other. *Hint:* start anywhere, and while someone has more enemies in their room than across the hall, move them; track the total number of enemy pairs sharing a room.
4. When a 5 × 5 board is covered by eight straight 1 × 3 tiles with one square left over, prove the uncovered square must be the *center*. *Hint:* color by $(x+y)$ mod 3 *and* by $(x-y)$ mod 3 to trap the hole, then exhibit a tiling that leaves the center empty.

THE FLASH *No clock — each cracks open in one flash of insight, once you find the angle. Nudges at the back.*

1. A knight makes 13 moves. Can it end on a square of the same color it started on?
2. Starting from $(0,0)$, each move adds $(\pm1,\pm1)$ to your position (you choose the signs each time). Can you ever land on $(0,1)$?

PROOF LAB *Write it clean, in full sentences.*

1. A circle has $n \geq 3$ lamps, all off. A move toggles a lamp and its two neighbors. Determine, with proof, for which n you can turn all lamps on.

2. Integers sit on a board. A move picks two of them, a and b with $a \geq b + 2$, and replaces them by $a - 1$ and $b + 1$ (shifting one unit from the larger to the smaller). Prove the process must terminate, by exhibiting a suitable monovariant.

3. An equilateral triangle of side n is cut into n^2 unit triangles. Prove it can never be tiled by rhombi (each rhombus being one up-triangle joined to one down-triangle), by two-coloring up against down and counting.

REFLECTION *No nudge. Write a few sentences.*

1. For one impossibility problem above, what *tipped you off* to which invariant to use? Was it the move's arithmetic, a coloring, or a parity? Record the cue for next time.

NUDGES

Launch. (1) No — position parity flips each jump; 7 odd lands at odd distance, never 0. (2) OFF — each toggle flips the state, and 100 is even, so it returns to its starting OFF. **Core.** (3) Each step drops the count of numbers by one and the quantity $\sum(\text{number}) - (\text{count})$ is invariant; the survivor is $1 + \cdots + 7 - 6 = 22$. (4) 25 squares is odd; a closed rook tour alternates colors and needs an even number of squares. (5) Track the parity of the white-stone count: every move changes it by an even amount, so it stays even (it starts at 8). The lone survivor can't be a single odd white, so it must be black. **Stretch.** (6) One swap is an odd permutation while the blank returns to its corner (even displacement) — the combined invariant differs from solved. (7) Write the final sign of each cell as $(-1)^{r_i+c_j}$ from row/column flip counts; the requirements at the center and the rest clash mod 2, so all +1 is unreachable. (8) Track the number of enemy pairs sharing a room; moving an unhappy person strictly lowers it, and a non-negative integer can't fall forever, so the process halts at a seating where nobody can improve. (9) The $(x + y)$ mod 3 coloring traps the hole in a 9-cell surplus class; the second coloring $(x - y)$ mod 3 traps it in another; the two classes meet *only* at the center $(3, 3)$, and an explicit tiling leaves the center — so the center is the only possibility. **The Flash.** (10) No — 13 odd flips color. (11) No — each move changes $x + y$ by an even amount, so its parity is locked at even, while $(0, 1)$ has odd coordinate-sum. **Proof.** (12) Solvable for *every* $n \geq 3$. Apply the move once at each of the n lamps; since toggles commute, only the number of flips per lamp matters, and a given lamp is toggled by the move at itself and by the moves at its two neighbors — exactly 3 times, which is odd. So every lamp finishes ON, and no value of n is obstructed. (13) Use the monovariant $S = \sum a_i^2$. Replacing a, b (with $a \geq b + 2$) by $a - 1, b + 1$ changes S by $(a - 1)^2 + (b + 1)^2 - a^2 - b^2 = -2(a - b - 1) \leq -2$, since $a - b - 1 \geq 1$, while the total $\sum a_i$ is unchanged. Thus S is a non-negative integer that strictly decreases at every move, so only finitely many moves are possible and the process halts — necessarily once all values lie within 1 of each other. (14) Two-color the cells by orientation: a side-n triangle holds $\frac{n(n+1)}{2}$ upward unit triangles and $\frac{n(n-1)}{2}$ downward ones, a surplus of n upward. Each rhombus covers exactly one upward and one downward cell, so any tiling would use equally many of each; since the counts differ by $n \geq 1$, no tiling exists.

CHAPTER FOUR 4

Count Something Else

Don't count the hard thing. Find an easy thing it secretly equals — pair them up perfectly — and count that instead.

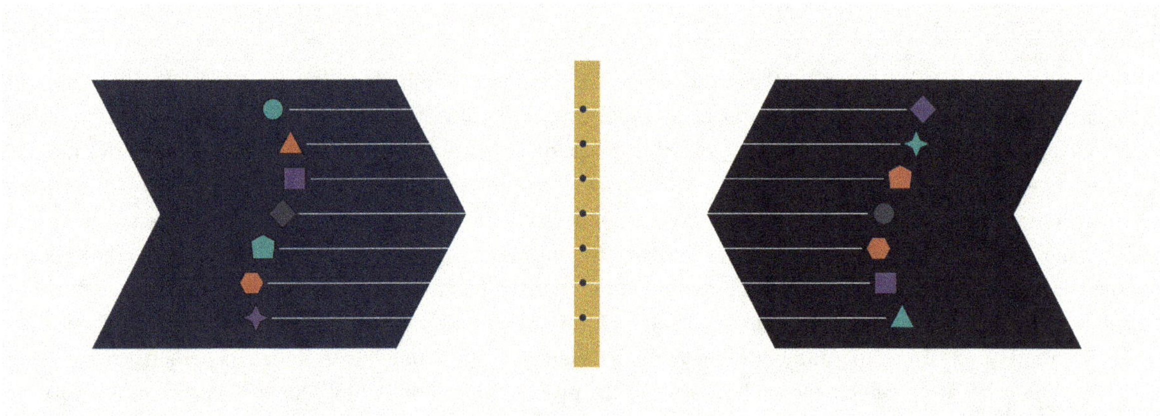

Bijections: don't count it — pair it.

Here's a question about subsets that seems to require careful bookkeeping.

WONDER THIS

Take a set of n things. Some of its subsets have an *even* number of elements (including the empty set, with zero); others have an *odd* number.

Which kind is more common — even or odd? Count the even-sized subsets and the odd-sized subsets. How do the two totals compare?

CURIOUS CARL • FIRST GUESS

Carl: "Two odd-looking sets, same size or not? I'll count each one and compare the totals. Numbers don't lie."

The catch: often there's no clean formula for either side, Carl — but a single one-to-one *pairing* proves the sizes match without your ever counting a thing.

You could add up binomial coefficients (the even-sized subsets number $\binom{n}{0} + \binom{n}{2} + \binom{n}{4} + \cdots$, the odd ones $\binom{n}{1} + \binom{n}{3} + \cdots$) and try to show the two sums are equal. That works, but it's algebra, and it doesn't really tell you *why*. There's a way to see the answer so clearly that no computation is needed at all. This is the art of translation: if an object is hard, trade it for an easy record of itself. Change the object, keep the answer — and in the right language, the hard question reads itself aloud.

Pick one specific element — call it element #1. Now take *any* subset and apply a single simple rule: *if element* #1 *is in it, remove it; if it's not, add it.* This "toggle #1" operation does two things. First, it changes the subset's size by exactly one, so it turns every even-sized subset into an odd-sized one and vice versa. Second, doing it *twice* brings you right back where you started — so it pairs each subset with exactly one partner, and that partner's partner is you again.

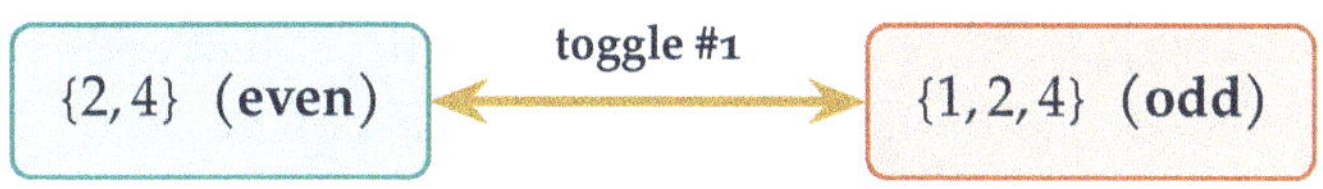

a perfect partner for every subset

So every subset is matched, one-to-one, with a subset of the opposite parity. The even-sized subsets and the odd-sized subsets are paired off into perfect couples — which means there must be *exactly the same number of each.* **For every $n \geq 1$, even and odd subsets are exactly tied,** 2^{n-1} of each, and we never added a single binomial coefficient. We just found the right pairing. (The lone exception is $n = 0$: the empty set is even, with no odd subset to partner: there's no element to toggle.)

That pairing is a **bijection**: a perfect one-to-one matching between two collections. And it carries a simple, powerful consequence.

THE MOVE

A **bijection** between two sets A and B is a pairing that matches every element of A with exactly one element of B, and vice versa, leaving nothing unpaired on either side. If such a pairing exists, then *A and B have the same size* — no matter how different they look. To count a hard set, find a bijection to an easy one. The whole skill is *building the matching.*

The toggle we just used has a name worth knowing: it is an *involution* — a pairing that is its own undo, so that doing it twice returns you to the start. Involutions are the sharpest instrument in this whole chapter. Because an involution with no fixed points splits a set into disjoint couples, it proves two families are equinumerous in a single stroke: one rule, no counting, no formula. (And when an involution does leave a few elements fixed — paired with themselves — those fixed points are often exactly the quantity a harder problem is really asking about.) Whenever you must show two counts are equal, the first thing to try is: *is there a do-it-twice-and-return pairing between them?*

This is counting's greatest sleight of hand. You never count the thing you were asked about. You find something it perfectly mirrors, and count *that.* Done well, a bijection turns a problem you can't do into one you already finished.

Encoding: when objects are secretly strings

The simplest bijections turn awkward objects into tidy ones (usually sequences or strings) that are trivial to count. The art is finding an *encoding* that loses no information, so the matching is truly one-to-one.

WATCH THIS — #1 Subsets are just yes/no lists

Claim. A set of n elements has exactly 2^n subsets.

The bijection. Match each subset with a string of n bits — one bit per element, reading 1 if that element is in the subset and 0 if it isn't.

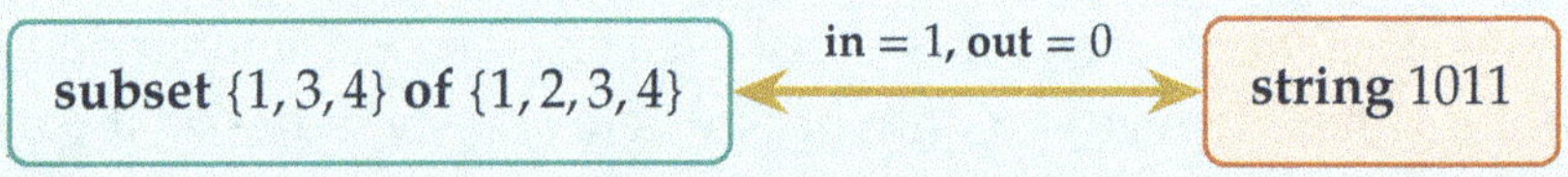

The count. This matching is perfectly reversible: every subset gives one string, every string gives back one subset, nothing is left over. So there are exactly as many subsets as there are length-n bit strings. And bit strings are easy: each of the n positions is independently 0 or 1, giving $2 \times 2 \times \cdots \times 2 = 2^n$. ■

Trigger A set of complicated objects you can't directly count, but which each carry a fixed amount of yes/no or positional information.

Do first Encode each object as a string or sequence, one symbol per "decision," making sure you can decode it back uniquely.

Trapdoor An encoding that two different objects can share, or that no object produces — then it isn't a bijection and the counts won't match.

Upgrade Once objects are strings, multiplication and known formulas finish instantly — the encoding is the whole battle.

That "2^n subsets" fact you may have memorized is, underneath, a bijection to bit strings. This is the recurring rhythm: a bijection lets a known, easy count stand in for a mysterious one. The cleverness lives entirely in the *matching*. Once it's built, the arithmetic is something you already know how to do.

Stars and bars: a bijection that builds itself

Some bijections are so good they've earned names. Here's the most useful counting bijection in all of combinatorics — the one that answers "how many ways can I distribute identical things into distinct boxes."

WATCH THIS — #2 Stars and bars

Claim. The number of ways to write n as an ordered sum of k *non-negative* whole numbers — that is, solutions to $x_1 + x_2 + \cdots + x_k = n$ — is $\binom{n+k-1}{k-1}$.

The bijection. Line up the n identical *stars* and $k-1$ *bars* in a single row, in any order — the bars may sit side by side or at either end. The bars cut the row into k compartments; the number of stars in each compartment gives $x_1, x_2, \ldots, x_k$ (two adjacent bars, or a bar at an end, mean that box gets 0).

★ ★ | ★ ★ ★ | ★ ★ → $(2,3,2)$, **a solution to** $x_1 + x_2 + x_3 = 7$

The count. Every arrangement of n stars and $k-1$ bars in a row encodes exactly one solution, and every solution gives exactly one arrangement — a clean bijection. So we just need to count arrangements of $n + (k-1)$ symbols where we choose which $k-1$ of the positions are bars:

$$\binom{n+k-1}{k-1}.$$

For $x_1 + x_2 + x_3 = 7$: that's $\binom{9}{2} = 36$ solutions. ■

Look at what the picture did: a question about *numbers* (sums equaling 7) became a question about *arrangements* (stars and bars in a row), and arrangements we count with a single binomial coefficient. We didn't solve any equations. We found the objects the solutions secretly are. Whenever you face "distribute n identical items into k labeled groups," reach for stars and bars; it has likely already done your work for you.

△ **TRAPDOOR**

A matching is only a bijection if it works *both ways*: every object on the left hits exactly one on the right, *and* every object on the right is hit exactly once. If two different objects map to the same target (not injective) or some target is never reached (not surjective), the counts won't be equal. Always check both directions before trusting a bijection — a one-way map is just a hopeful guess.

CURIOUS CARL • TRAPDOOR!

Carl pairs each subset of $\{1,2,3\}$ with its *smallest* element and declares, "There — a perfect matching, so the counts must be equal!"

The empty set raises a hand: *"I have no smallest element, so nobody's holding my hand — and* $\{1\}$, $\{1,2\}$, *and* $\{1,3\}$ *all grabbed the same target,* 1." A map that misses one target and double-covers another isn't a bijection at all, Carl — check that it's reversible *onto* the whole right-hand side before you trust the count.

The showpiece: the Catalan numbers

Now for a bijection so rich that the same count appears in dozens of disguises across mathematics — a number sequence that shows up whenever something has to stay "balanced."

WONDER THIS

You're at a counter selling tickets for \$1. Customers arrive one at a time: n of them carry only a \$1 bill, and n of them carry only a \$2 bill (and want \$1 change). You start with an empty till.

In how many arrival orders can you *always* make change — never caught needing to hand out a dollar you don't yet have?

This looks like it needs careful case-tracking. But it's one of the most famous counting problems in mathematics in disguise, and a bijection cracks it wide open.

WATCH THIS **#3 Balanced paths and the reflection trick**

Reframe as a path. Write a $+1$ when a \$1 customer arrives (the till gains a dollar) and a -1 when a \$2 customer arrives (the till must give one back). Reading left to right, you can always make change exactly when every running total stays ≥ 0. Picture it as a path that steps up on $+1$ and down on -1: it must *never dip below the floor.*

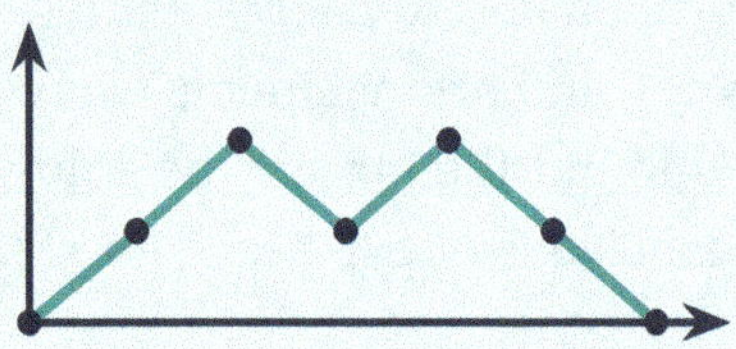

stays on or above the floor

The count. The total number of arrival orders, ignoring the rule, is $\binom{2n}{n}$ — choose which n of the $2n$ slots hold the \$1 customers. The *bad* orders are the ones whose path dips to -1 at some point. Here's the magic, called the *reflection principle*: take a bad path and, from the first moment it touches -1, flip every later step up-for-down. This reflection is a perfect bijection between bad paths and *all* paths ending at a different height — and those are easy to count: there are $\binom{2n}{n+1}$ of them.

The finish. Good orders are all orders minus bad orders:

$$\binom{2n}{n} - \binom{2n}{n+1} = \frac{1}{n+1}\binom{2n}{n} =: C_n,$$

the n-th **Catalan number**. For small n these are $1, 1, 2, 5, 14, 42, \ldots$ — so with 3 customers of each type, there are $C_3 = 5$ safe arrival orders. ■

THE TURN *from grind to insight*

The brute-force way to count the safe orders is to count them *directly* — march through arrangements, track the running till, and tally the ones that never go negative. The bad cases tangle together, there is no obvious formula to aim for, and you would be bookkeeping forever.

The reflection trick is to never count the good paths at all. Instead, take each *bad* path and flip its tail at the first dip below the floor — a single reflection that pairs every bad path, one-to-one, with a path in a family you *can* count on sight. The matching does the counting: subtract the easy family from the total and the Catalan number drops out. When a set resists direct counting, don't push harder — find the perfect pairing that turns it into a set you already know.

The reflection trick is one of the most beautiful bijections in all of mathematics: it matches the hard-to-count "bad" paths with an easy-to-count family, just by flipping a tail. And the punchline is that C_n is everywhere. The same Catalan numbers count the ways to correctly nest n pairs of parentheses, the ways to triangulate a polygon, the ways to build a binary tree with n nodes, the paths through a grid that stay below the diagonal. All the *same count*, because there are bijections quietly linking every one of these to our ticket-counter paths. Find one Catalan structure and you've found them all.

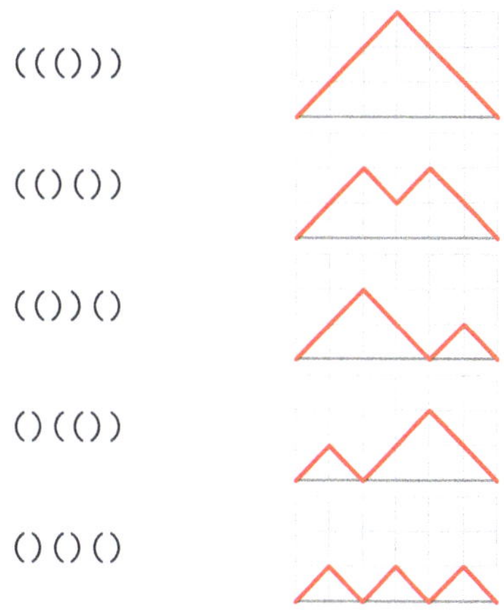

$$C_3 = 5 = \frac{1}{4}\binom{6}{3}$$

$C_3 = 5$: the five balanced parenthesizations of three pairs and their matching Dyck paths.

SEEN BEFORE?

The Catalan numbers will return as the stars of **Chapter 5** (Recursion), where the same count falls out of a recurrence instead of a bijection, and again in **Chapter 8** (Generating Functions), where a single equation produces them in one stroke. Three completely different routes, one sequence — a preview of how the chapters secretly talk to each other.

TRY TO BREAK IT

A real bijection has to survive being run *backwards*. So try to break a plausible one: hunt for a map between two sets that looks one-to-one but secretly misses a target or covers one twice. For instance, "pair each up/down path with the position of its *first* down-step" feels tidy — until two different paths share that first down-step, or the all-up path has none at all. When a candidate map fails one direction, it proves only an inequality, never equality; feeling exactly where it snaps is how you learn to trust the ones that hold.

WATCH THIS **#4 The Catalan zoo: one count, many costumes**

Claim. Balanced parentheses, Dyck paths, and non-crossing handshakes are all counted by C_n — because explicit bijections tie them together.

Parentheses ↔ paths. Read each "(" as an up-step and each ")" as a down-step. A string of n pairs is *balanced* exactly when the path never dips below the floor — the Dyck paths we just counted.

Parentheses ↔ handshakes. Seat $2n$ people around a table and pair them off with n handshakes so that *no two chords cross*. Join each "(" to its matching ")" and connect those two seats: properly nested parentheses correspond exactly to non-crossing chords.

(()) ↔ three non-crossing chords

So all three families have the same size, C_n: nest n pairs of parentheses and you have *simultaneously* drawn a mountain range and a non-crossing handshake. **One count, wearing every costume in the zoo.** ■

Payoff: cutting a number into pieces

One more, fast, now that you can spot a bijection coming.

WONDER THIS

A *composition* of a number is a way of writing it as an ordered sum of positive parts: $4 = 1 + 3$ and $4 = 3 + 1$ count as *different* compositions.

How many compositions does the number n have?

Picture n identical dots in a row. Between consecutive dots there are $n - 1$ gaps, and into each gap you may either drop a "cut" or not. Each pattern of cuts slices the row into consecutive runs — and the run lengths are exactly the parts of a composition.

• / • • / • → $1 + 2 + 1$, **a composition of** 4

Each of the $n - 1$ gaps is an independent yes/no choice — cut or don't — so there are 2^{n-1} patterns, hence 2^{n-1} compositions of n. For $n = 5$ that's $2^4 = 16$. **The gaps were the bijection all along**: compositions of n correspond perfectly to subsets of the $n - 1$ gaps, and we already know a set of $n - 1$ things has 2^{n-1} subsets. *We've now counted the same 2^{n-1} three different ways in one book. It keeps showing up because it keeps being right.*

Partitions: when order stops mattering

Compositions counted *ordered* sums. Drop the order (treat $1 + 3$ and $3 + 1$ as the same) and you get a *partition*, one of the richest objects in combinatorics. Partitions have no simple closed-form count, yet bijections still expose their hidden symmetries with almost no computation.

A partition of n writes it as a sum of positive parts where only the *multiset* of parts matters. We draw one as a *Ferrers diagram*: a left-justified array of dots, one row per part, longest on top.

Conjugation: a free bijection. Reflect a Ferrers diagram across its main diagonal — rows become columns. The result is another valid partition of n, its *conjugate*, and conjugation is a perfect self-reversing pairing (flip twice, you're back). Read off one consequence instantly: **the number of partitions of n into exactly k parts equals the number whose largest part is k**, because the flip turns "number of rows" into "length of the top row." One reflection, a theorem.

WATCH THIS — #5 Euler's theorem: distinct parts vs. odd parts

Claim. The number of partitions of n into *distinct* parts equals the number into *odd* parts (repeats allowed).

For $n = 6$: the distinct-part partitions are 6, 5+1, 4+2, 3+2+1 — four of them. The odd-part partitions are 5+1, 3+3, 3+1+1+1, 1+1+1+1+1+1 — also four. The match is no accident.

The bijection. Every positive integer is (an odd number) × (a power of 2) in exactly one way. Pair the two sides with that fact:

- **Distinct → odd.** Take a distinct part, write it as odd × 2^a, and replace it with 2^a copies of that odd number. (E.g. $12 = 3 \times 4 \to 3{+}3{+}3{+}3$.)
- **Odd → distinct.** Take an odd value appearing c times; write c in binary, $c = 2^{a_1} + 2^{a_2} + \cdots$, and merge those copies into the distinct parts odd × 2^{a_1}, odd × 2^{a_2}, ...

Binary's uniqueness makes each step reversible, so it's a genuine bijection — the two counts must be equal, for every n at once. ■

There is no tidy formula for $p(n)$, the total number of partitions: it grows wildly ($p(10) = 42$, and $p(100)$ exceeds 190 million). Yet bijections still pin down exact *relationships* between kinds of partitions with no arithmetic at all. That is the bijective principle at its most powerful: *equality without enumeration.*

Counting trees: Prüfer's sequence and Cayley's formula

How many different trees can you build on a fixed set of labeled dots? The answer is astonishingly clean. And, once again, a bijection to plain sequences proves it.

WONDER THIS

Label n dots $1, 2, \ldots, n$. A *tree* joins them all with $n-1$ edges and no cycle.

How many different labeled trees are there? For $n = 3$ there are 3; for $n = 4$ there are 16; for $n = 5$, 125. **What is the pattern?**

Those counts 3, 16, 125 are $3^1, 4^2, 5^3$, suspiciously like n^{n-2}. Checking cases would be hopeless, but a bijection to short strings makes it fall out.

A good bijection is a *lossless file format* for a mathematical object: it must encode the object so that *nothing* is thrown away and the original can be rebuilt exactly. That is a real design constraint, and lazy encodings fail it. You might try to encode a tree by its *list of vertex degrees* — but that is lossy: the two different trees on $\{1, 2, 3, 4\}$ given by the path 1–2–3–4 and the path 1–3–2–4 both have degree list 1, 2, 2, 1, so the degrees cannot tell them apart. The Prüfer sequence is the encoding that keeps *exactly* the missing information: *which* vertex each leaf hangs from, in the right order.

WATCH THIS — **#6 Cayley's formula via Prüfer codes**

Claim (Cayley). There are exactly n^{n-2} labeled trees on n vertices.

Encode a tree as a sequence. Given a labeled tree with $n \geq 2$ vertices, repeat $n-2$ times: locate the *leaf* (a degree-1 vertex) with the smallest label, write down the label of its single neighbor, then delete that leaf. The record is a *Prüfer sequence* of length $n-2$, each entry a label between 1 and n.

Decode back. Run the encoding in reverse, keeping the set of labels not yet placed. At each step, read the *first* entry s still in the sequence, find the smallest label ℓ that does *not* appear anywhere in the remaining sequence, join ℓ to s, then delete ℓ from the label set and drop that first entry from the sequence. When the sequence is finally empty, exactly two labels remain — join them. Each step is forced: a label appears in the sequence exactly (its degree -1) times, so the labels *absent* from the remaining sequence are precisely the current leaves, and the smallest such leaf is the one the encoder removed. Encoding and decoding are therefore exact inverses — a genuine bijection between trees and sequences.

The count. A Prüfer sequence is just a string of length $n-2$ over an alphabet of n labels, and there are n^{n-2} such strings. So there are exactly n^{n-2} labeled trees. For $n = 4$: $4^2 = 16$. ∎

A tree (a tangled, branching, two-dimensional object) turned out to be nothing more than a short list of numbers in disguise. That is the encoding move from the top of the chapter aimed at a far wilder target: subsets became bit strings; now trees become Prüfer strings. *Whenever a structure has a forced "smallest" piece you can strip away and record, look for the sequence hiding inside it.*

TRY TO BREAK IT

The Prüfer encoding needs $n \geq 2$ (with $n - 2 \geq 0$), and the "smallest leaf" step needs there to *always* be at least two leaves until one vertex remains. Try to break it: can a tree on ≥ 2 vertices ever have fewer than two leaves? Follow the longest path to its ends and you will find two leaves waiting — which is exactly why the algorithm never gets stuck.

When counting proves a theorem: Fermat by necklaces

Bijections don't only count. They can *prove*. Here a pure counting argument settles a landmark of number theory using no number theory at all.

WONDER THIS

Fix a prime p and a colors. Thread p beads on a circle, each bead any of the a colors.

How many genuinely different necklaces result — counting two as the same when one is a rotation of the other? Track the count carefully and a classic theorem falls out.

WATCH THIS **#7 Fermat's little theorem, counted**

Setup. There are a^p colored *strings* of p beads (each bead independently one of a colors). Call two strings equivalent when one is a rotation of the other — *rotations only*, we never flip the necklace over — and each equivalence class is a *rotational necklace.*

The key fact (p prime). A string is left unchanged by some nontrivial rotation only if all its beads are identical — and there are exactly a such monochromatic strings. Every *other* string has p *distinct* rotations, because p is prime: no shift by $1, 2, \ldots, p-1$ can bring a non-constant string back to itself. So the non-monochromatic strings fall into groups of *exactly* p.

Count two ways. The a^p strings are the a monochromatic ones plus the rest, and the rest come in blocks of p:

$$a^p = a + p \cdot (\text{number of multicolor necklaces}).$$

Hence $p \mid a^p - a$ — which is **Fermat's little theorem,** $a^p \equiv a \pmod{p}$. ∎

Not one congruence was manipulated. We counted a set of strings, noticed that rotation sorts them into blocks of p (with a monochromatic leftovers), and the divisibility *was* the theorem. This is a *group action* counting argument in miniature: rotations act on the strings, and because p is prime, every orbit has size 1 or p.

SEEN BEFORE?

Fermat's little theorem is the engine of **Chapter 10**, proved there by modular algebra — yet here the very same theorem drops out of counting beads. And "orbits have size 1 or p" is the first taste of **Chapter 16** (Symmetry), where counting configurations *up to rotation* becomes a general machine.

What to try when you're stuck

When you're asked "how many" and a direct count looks painful — or when you must show two collections are the same size — think bijection:

1. **Name the hard set** you're trying to count.

2. **Find an easy set** you already know how to count: strings, subsets, lattice paths, arrangements.
3. **Build a matching** between them — a rule turning each hard object into one easy object.
4. **Check both directions.** Can you reverse the rule and recover the original uniquely? If yes, it's a bijection and the sizes are equal.
5. **For pairing arguments,** a self-inverse matching (do it twice, return to start) proves two halves are equal — like the parity-toggle on subsets.
6. **For exotic objects** (trees, or anything with a forced "smallest" feature), strip that feature repeatedly and record it, encoding the object as a short sequence — as Prüfer does for trees.
7. **To prove a divisibility or identity,** count one set two ways, or count it *up to symmetry* so equal-size orbits force the arithmetic — as necklaces prove Fermat.

THE MOVE

One-page recap. A bijection is a perfect one-to-one matching; if one exists, the two sets are the same size. Don't count the hard thing — count an easy thing it mirrors. Toolbox: **encode as strings** (subsets ↔ bit strings), **stars and bars** (identical items into labeled boxes), **reflection** (Catalan / paths that stay above a floor), **Ferrers conjugation** (flip a partition diagonally to swap rows and columns), **toggle / involution** (pairing to prove two counts equal), **Prüfer codes** (a tree becomes a short string, giving Cayley's n^{n-2}), and **counting up to rotation** (necklaces prove Fermat's little theorem). When stuck, ask: *what easy object is this secretly?*

THE NAME FOR IT

Bijections

A perfect one-to-one pairing between two sets is a *bijection*, and two sets joined by one must have equal size — the *bijective principle.* The recurring counts here, $1, 1, 2, 5, 14, 42, \ldots$, are the *Catalan numbers;* the path-flipping trick is the *reflection principle;* and writing n as an unordered sum of parts is an *integer partition,* with Euler's distinct-equals-odd theorem its first jewel.

WHERE IT LEADS A bijection is the foundation of *cardinality* — Cantor's discovery that infinite sets come in different sizes rests entirely on which bijections exist. Counting one structure by matching it to another is the engine of *enumerative combinatorics* and the *species* and *bijective-proof* programs of modern algebra, and the theory of *integer partitions* blooms into the *Rogers–Ramanujan identities* and the modular forms of analytic number theory.

CURIOUS CARL • EUREKA

Carl watches every squid pair off with exactly one teapot. "I was about to count both crowds. *Twice.* They just reached across and settled it themselves — nobody needed me at all."

SAME BEAST, NEW COSTUME

The same "pair it, don't count it" beast, in costume across the book:

Subset ↔ bit-string → Ballot paths ↔ parentheses → Catalan objects → Ferrers conjugation → Trees ↔ Prüfer strings → Necklaces up to rotation.

Each one quietly matches a hard set to an easy one, element for element, and lets the easy count stand in. Build the matching and the sizes must agree.

The practice ladder

THE PRACTICE LADDER *climb at your own pace*

LAUNCH *Warm the engine. Nudges below.*

1. How many ways can 5 identical candies be given to 3 children (some may get none)? Use stars and bars.
2. Show that a set of n elements has the same number of subsets of size k as of size $n-k$, by an explicit matching.

CORE BUILDER *Master the central moves.*

1. How many monotone lattice paths go from $(0,0)$ to (m,n) using only unit steps right and up? Find the bijection to strings and count.
2. For $n \geq 1$, prove the number of subsets of $\{1,\dots,n\}$ with an even number of elements equals the number with an odd number, using the toggle bijection.
3. How many ways can you write n as an ordered sum of k *positive* parts? (Adapt stars and bars.)
4. Using Ferrers conjugation, show that the number of partitions of n into at most k parts equals the number into parts each at most k.

STRETCH *Transfer to unfamiliar ground.*

1. Show that the number of ways to correctly nest n pairs of parentheses equals C_n, by a bijection to balanced up/down paths.
2. Prove $\sum_{k=0}^{n}\binom{n}{k}^2=\binom{2n}{n}$ by counting lattice paths through a midpoint two ways.

THE FLASH *No clock — each cracks open in one flash of insight, once you find the angle. Nudges at the back.*

1. How many solutions in non-negative integers does $a+b+c+d=10$ have?
2. In how many ways can 6 people around a round table be split into pairs of adjacent neighbors? (Reframe as a path or matching.)

PROOF LAB *Write it clean, in full sentences.*

1. Prove the reflection principle: the number of up/down paths from height 0 that touch -1 equals the number of *all* paths from height -2 to the same endpoint. Use it to derive $C_n = \binom{2n}{n} - \binom{2n}{n+1}$.

2. Give a bijection between binary trees with n internal nodes and triangulations of a convex $(n+2)$-gon, and conclude both are counted by C_n.

3. Prove Euler's theorem that partitions of n into distinct parts are equinumerous with partitions into odd parts, by carefully justifying that the odd $\times 2^a$ factorization makes the stated map a bijection in both directions.

4. Give an explicit bijection between balanced strings of n pairs of parentheses and non-crossing perfect matchings ("handshakes") of $2n$ points on a circle, and conclude both are counted by C_n.

5. Using Prüfer sequences, prove that the number of labeled trees on n vertices is n^{n-2}, and illustrate the encoding on one tree with 4 vertices.

6. By counting colored strings of p beads up to rotation (p prime), prove Fermat's little theorem $a^p \equiv a \pmod{p}$.

REFLECTION *No nudge. Write a few sentences.*

1. Which bijection above felt least obvious to invent? Describe the moment you saw what the hard objects "really were." That recognition is the skill — name its trigger.

NUDGES

Launch. (1) $\binom{5+2}{2} = \binom{7}{2} = 21$. (2) Match each k-subset with its complement, an $(n-k)$-subset — reversible, hence a bijection. **Core.** (3) Encode a path as a string of m R's and n U's; count $\binom{m+n}{m}$. (4) Toggle element #1; it flips parity and is self-inverse, pairing the two families. (5) Give each part one unit first, then stars-and-bars the rest: $\binom{n-1}{k-1}$. (6) Conjugation swaps "number of parts" (rows) with "size of largest part" (top row length), so "at most k parts" ↔ "largest part at most k." **Stretch.** (7) Match "(" with an up-step and ")" with a down-step; valid nesting ↔ path staying ≥ 0. (8) A path to (n,n) through the midpoint splits as $\binom{n}{k}\binom{n}{n-k}$; sum over k. **The Flash.** (9) $\binom{10+3}{3} = \binom{13}{3} = 286$. (10) Reframe as tilings of a cycle; count the valid adjacent-pairings. **Proof.** (11) Reflect a path across the line $y = -1$ starting at its first touch of -1: this sends the initial height 0 to -2, fixes the endpoint, and is its own inverse, so it bijects paths from 0 that reach -1 with all paths from -2 to the same endpoint. To count $2n$-step paths from 0 to 0 that never dip below 0, subtract the "bad" ones (those touching -1) from all $\binom{2n}{n}$; the bad ones correspond to paths from -2 to 0, of which there are $\binom{2n}{n+1}$, giving $C_n = \binom{2n}{n} - \binom{2n}{n+1}$. (12) Fix one edge of the $(n+2)$-gon; in any triangulation exactly one triangle sits on it, and its third vertex splits the polygon into a left and a right sub-polygon. Send that triangle to the root of a binary tree and recurse on the two pieces for the left and right subtrees — a reversible step, hence a bijection. Both sides satisfy $X_n = \sum_i X_i X_{n-1-i}$ with $X_0 = 1$, the Catalan recurrence, so triangulations and binary trees are both counted by C_n. (13) Map a distinct-part partition to an odd-part one by writing each part uniquely as (odd) $\times 2^a$ and replacing it with 2^a copies of that odd core. Reverse it by taking each odd value v used c times, writing c in binary $c = \sum 2^{a_i}$, and merging into the parts $v \cdot 2^{a_i}$ — distinct both within v and across different odd cores. The two maps preserve the sum and are mutually inverse, so the two kinds of partition are equinumerous. (14) Place the $2n$ symbols on a circle and join each "(" to the ")" that closes it by a chord; proper nesting means no two chords cross, and every non-crossing matching reads back a balanced string, so this is a bijection. Balanced strings obey the Catalan recurrence, hence both families number C_n. (15) Form the Prüfer code by repeatedly deleting the smallest-labeled leaf and recording its unique neighbor, stopping after $n-2$ deletions, which yields a string in $\{1,\dots,n\}^{n-2}$. Each vertex appears exactly $\deg -1$ times, and that rule lets you rebuild the tree uniquely, so labeled trees biject with these strings and number n^{n-2}; for instance the star on $\{1,2,3,4\}$ with center 1 codes as $(1,1)$ — strip leaf 2 then leaf 3, each with neighbor 1, the center's degree 3 showing as its two appearances. (16) Form all a^p strings of p beads in a colors and let the rotations act. Since p is prime, any string fixed by a nontrivial rotation is monochromatic, so the a constant strings are the only fixed points and every other string lies in an orbit of size exactly p; hence $p \mid a^p - a$, i.e. $a^p \equiv a \pmod{p}$.

MATH LAB #1

Build a Proof Machine

Four chapters, four master moves. Snap them together and aim them at fresh problems — then decide, each time, which gear the problem wants.

You've now met the first four moves of the book: **double counting** (one set, two stories), **pigeonhole** (too many objects, too few labels), **invariants** (watch what can't change), and **bijections** (don't count it — pair it). Each was powerful alone. Here you bolt all four onto one machine and feed it problems — the trick is choosing which gear engages.

CURIOUS CARL • FIRST GUESS

Carl: "A machine with four gears? I'll crank *all four* on every problem and see what pops out."

The catch: most problems want exactly *one* gear, Carl — the whole skill is recognizing which. Read each station and ask: is this secretly a *count*, a *squeeze*, a *frozen quantity*, or a *pairing*?

STATION 1 • the Double-Counting gear

In a school, every student belongs to exactly 3 clubs, and every club has exactly 5 members. There are 12 clubs. **How many students are there?**

Turn the gear: count the pairs (student, club-they're-in) two ways.

STATION 2 • the Pigeonhole gear

Five darts land inside a 2×2 square. **Prove two of them land within $\sqrt{2}$ of each other.**

Turn the gear: what four "holes" can you cut the square into, and how far apart can two points in one hole be?

STATION 3 • the Invariant gear

Fifteen cups sit on a table, all upside-down. A move flips *exactly two* cups at once (your choice which two). **Can you ever get all fifteen right-side-up?**

Turn the gear: track the *parity* of the number of right-side-up cups. How does one move change it?

STATION 4 • the Bijection gear

Prove that the set $\{1, 2, \ldots, n\}$ has exactly as many subsets of *even* size as of *odd* size.

Turn the gear: given any subset, toggle whether the element 1 is in it. What does that do to the size's parity — and is the move reversible?

STATION 5 • Boss problem — you choose the gear

Six people are at a party; friendship is mutual, and each person's number of friends *present* is somewhere from 0 to 5. **Prove that two people have the *same* number of friends present.**

Which gear? It looks like pigeonhole (6 people, 6 possible counts 0–5) — but 6 holes for 6 people isn't enough. Find the reason the values 0 and 5 can't *both* occur, and the pigeonhole snaps shut.

NUDGES

Peeks (try first!). **(1)** Memberships = $12 \times 5 = 60 = 3 \times$ (#students), so $\boxed{20}$ students. **(2)** Cut the square into four 1×1 cells; 5 darts, 4 cells force two into one cell, and two points in a unit square are at most its diagonal $\sqrt{2}$ apart. **(3)** No. Each flip changes the count of upright cups by $+2, 0,$ or -2 — always *even* — so its parity is frozen at even (it starts at 0). But 15 is odd, so all-upright is unreachable. **(4)** Toggling element 1 pairs each even subset with an odd one and is its own inverse — a bijection — so the two families are equal in size (each is 2^{n-1}). **(5)** If someone has 0 friends present, nobody has 5 (and vice versa), so at most 5 distinct values occur among 6 people — pigeonhole forces a repeat.

CURIOUS CARL • EUREKA

Carl: "Oh — the machine doesn't do the thinking. *Naming the move* is the thinking. Once I see 'this is really a count' or 'this is really a pairing,' the gear turns itself!"

Chapter Five

5

Solve It Smaller

Don't solve the whole problem. Solve a smaller copy of it, then let the small answers build the big one.

Recurrences: solve the smaller shadow.

Here's a counting question you can start by just...counting.

WONDER THIS

You have a $2 \times n$ strip of squares and a pile of 1×2 dominoes. Each domino covers two adjacent squares — placed either standing up or lying down.

In how many ways can you tile the whole strip? Try $n = 1, 2, 3, 4$ by hand first, and watch for a pattern.

CURIOUS CARL • FIRST GUESS

Carl: "A 2×20 strip? I'll *draw every tiling* and count the stack. Tedious, sure, but foolproof."

The catch: the tilings explode into the thousands, Carl — but look only at the *last column*, and each big strip quietly points back to a smaller one you have already solved.

Do the small cases. A 2×1 strip has just 1 tiling (one vertical domino). A 2×2 strip has 2 (two verticals side by side, or two horizontals stacked). A 2×3 has 3; a 2×4 has 5. So the counts run $1,2,3,5,\dots$, and if you know the Fibonacci numbers, your neck hairs should be standing up right now.

Why Fibonacci? Don't tile the whole strip — just think about the *very last column*. There are exactly two ways it can be covered. Either a single vertical domino fills it, which leaves a $2\times(n-1)$ strip to tile in every possible way; or the last column is covered by the right halves of two horizontal dominoes, whose left halves stick into the second-to-last column. That leaves a $2\times(n-2)$ strip. Those two scenarios don't overlap and they cover every tiling, so:

$$T_n = T_{n-1} + T_{n-2}.$$

The number of tilings of a $2\times n$ strip is a Fibonacci number. We never tiled a single large strip. We found how a big tiling *reduces* to smaller ones, and the reduction did all the work.

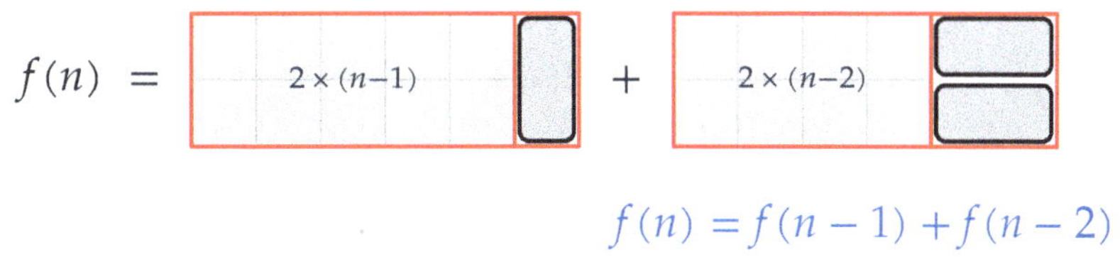

The last column of a $2\times n$ strip is either one vertical domino (leaving $2\times(n-1)$) or two horizontals (leaving $2\times(n-2)$), giving $f(n) = f(n-1) + f(n-2)$.

THE TURN *from grind to insight*

The head-on way to count the tilings is to start laying dominoes and tally arrangements — but the cases explode, and for a 2×20 strip you would be at it all day and still miscount. **The shrink** is to refuse to tile the strip at all. Look only at the *last column*: it can be finished in just two ways, and each way hands you a *smaller* strip you have already solved. We never tiled a single large strip; we found how a big one collapses into small ones, and the collapse did all the counting. When a problem is too big to attack head-on, don't — ask how it shrinks.

That equation, $T_n = T_{n-1} + T_{n-2}$, is a **recurrence**: a rule that defines each value in terms of earlier ones. The move this chapter teaches is to stop attacking a problem head-on and ask instead how it shrinks.

THE MOVE

To set up a **recurrence**, ask: "*How does a size-n problem break into smaller copies of the same problem?*" Find a single feature to split on — the last column, the largest element, the first move, the top disk — so that every large case falls into one of a few smaller cases with no overlaps and no gaps. The recurrence plus a few starting values *is* the solution. The art is choosing what to split on.

A recurrence trades one hard question for the relationship between easier ones. You don't need a formula to *use* a recurrence. You can compute as far as you like by building up. And often, the recurrence reveals more about *why* an answer is what it is than any closed formula ever would.

A recurrence is not a formula; it is a choice of *state*: of exactly what the future needs to know about the past. That is the real creative act, and it is usually a compression: the domino tilings need only the *last two* counts, never the whole history; the Josephus survivor needs only the circle's *size after one full lap*; the power sums coming later need only the *previous two* values. Pick the smallest packet of information that determines the next step, and the recurrence writes itself. Pick too little and the sequence isn't even determined. A two-term rule with a single seed can still finish a dozen different ways.

Splitting on the first move

The tiling recurrence came from looking at the last column. The same instinct (isolate one decision, then recurse on what's left) cracks a famous puzzle about moving disks.

WATCH THIS **#1 The Tower of Hanoi**

The puzzle. Three pegs; n disks of distinct sizes stacked largest-to-smallest on the first peg. Move the whole stack to the third peg, one disk at a time, never placing a larger disk on a smaller one. How many moves does it take?

Split on the biggest disk. The largest disk can only move once you've gotten the other $n-1$ disks out of its way — entirely onto the spare peg. So any solution must:

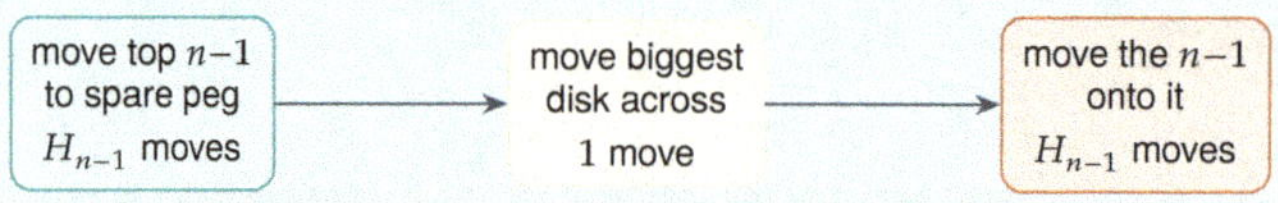

The recurrence. Reading the three stages, $H_n = 2H_{n-1} + 1$, with $H_1 = 1$.
Solving it. Compute upward: $1, 3, 7, 15, 31, \ldots$ — each one is one less than a power of two. Indeed $H_n = 2^n - 1$, which you can confirm by checking it satisfies the recurrence: $2(2^{n-1}-1)+1 = 2^n-1$. ■

Trigger A process built from a single repeatable action where each stage exposes a smaller copy of the same task.

Do first Isolate the one move that must happen (the biggest disk crossing), and recurse on the sub-tasks flanking it.

Trapdoor Forgetting that the sub-task happens *twice* (clear the disks, then restack them) — that doubling is the heart of the $2H_{n-1}$.

Upgrade The same "do it, flip the key piece, undo it" structure appears in many move-counting problems; once you see the recurrence, the closed form often follows by spotting the pattern.

Notice the shape of the argument: we never planned out all $2^n - 1$ moves. We described how an n-disk solution *contains* two $(n-1)$-disk solutions, and the count followed. This is recursion's defining pleasure: you solve a problem by assuming you've already solved its smaller version, then doing a little extra. It can feel almost like cheating. It isn't; it's just leverage.

Induction: making the pattern a proof

Every recurrence so far ended the same slightly suspicious way: we computed a few terms, *guessed* a closed form, and checked it against the recurrence. But checking five cases is not the same as proving infinitely many. Maybe the pattern that held for $n = 1, \dots, 5$ quietly breaks at $n = 600$. We need a way to pin down all infinitely many cases with a single finite argument. That way is *induction*. And it is the proof engine that every recurrence quietly runs on.

WONDER THIS

We guessed that the Tower of Hanoi takes $H_n = 2^n - 1$ moves, after watching $1, 3, 7, 15$ roll by.

How could a finite proof guarantee this for *every* n at once — including the cases far too large to ever check by hand?

The idea is to prove two small things and let them cascade. First, show the claim holds at the very bottom. Then show that *whenever* it holds at one size, it is forced to hold at the next. Those two facts together topple every case, like an infinite line of dominoes: knock over the first, guarantee each domino fells its neighbor, and the whole line must fall.

THE MOVE

Mathematical induction. To prove a statement $P(n)$ for every integer $n \geq n_0$:

- **Base case.** Check $P(n_0)$ directly.
- **Inductive step.** Assume $P(k)$ holds — the *induction hypothesis* — and use it to prove $P(k+1)$.

The base case starts the chain; the step propagates it forever. Together they establish $P(n)$ for all $n \geq n_0$ — infinitely many statements from two finite checks.

WATCH THIS #2 The Hanoi formula, finally proved

Claim. The recurrence $H_n = 2H_{n-1} + 1$ with $H_1 = 1$ has closed form $H_n = 2^n - 1$ for all $n \geq 1$.

Base case. $H_1 = 2^1 - 1 = 1$, matching the given seed. ✓

Inductive step. Assume $H_k = 2^k - 1$ for some $k \geq 1$. Then, using the recurrence and *only* the hypothesis,

$$H_{k+1} = 2H_k + 1 = 2(2^k - 1) + 1 = 2^{k+1} - 2 + 1 = 2^{k+1} - 1,$$

which is exactly the claim at $k + 1$. By induction, $H_n = 2^n - 1$ for every $n \geq 1$. ■

Trigger Any claim of the form "this formula or property holds for all n," especially one you found by guessing from small cases.

Do first Nail the base case, then assume the statement at k and grind out the statement at $k + 1$ using the recurrence or definition.

Trapdoor Skipping or fudging the base case. A flawless inductive step proves nothing if the chain never starts — exactly how the bogus "all horses are the same color" proof fails.

Upgrade When the step needs more than the immediately previous case, strengthen to *strong* induction: assume the claim for *all* smaller values at once.

Ordinary induction reaches back exactly one step: $P(k) \Rightarrow P(k+1)$. But some arguments need to lean on *several* earlier cases, or on *all* of them. For those, we simply assume more.

WATCH THIS **#3 Strong induction: every integer factors into primes**

Claim. Every integer $n \geq 2$ is a product of primes.

Base case. $n = 2$ is prime — a product of a single prime. ✓

Strong inductive step. Assume the claim for *every* integer from 2 up to k. Consider $k+1$. Either $k+1$ is prime, and we are done, or it factors as $k+1 = ab$ with $2 \leq a, b \leq k$. By the *strong* hypothesis, both a and b are products of primes; concatenating those factorizations writes $k+1$ as a product of primes too. By strong induction, every $n \geq 2$ factors into primes. ∎

The difference is the reach of the hypothesis: ordinary induction assumed the result at k alone; here we needed it for *all* values below $k+1$, because a and b could land anywhere down the line. **Strong induction** is the natural partner of recurrences that reach back more than one step. A two-term relation like Fibonacci leans on the *two* previous values, so proving things about it usually wants two base cases and a hypothesis covering both.

△ TRAPDOOR

Strong induction needs *enough* base cases. A step that uses $P(k)$ and $P(k-1)$ to reach $P(k+1)$ must verify *two* starting cases by hand; check only one and the very first step has nothing beneath it. The same warning applies to every two-term recurrence: one seed never determines the sequence, and one base case never anchors the induction.

SEEN BEFORE?

Induction is the constructive twin of the *extremal* and *descent* arguments of **Chapter 3**: "there is no smallest counterexample" is induction read backwards: the *well-ordering* principle. The same machinery drives the proofs of Fermat's and Euler's theorems in **Chapters 9–10**, and underwrites the closed forms we extract next.

Thinning the circle: the Josephus problem

A recurrence need not shrink a problem by a single step. Sometimes the reduction cuts the size in *half*. And a puzzle that looks hopeless to simulate folds into two lines.

WONDER THIS

n people stand in a circle, numbered 1 to n clockwise. Starting the count at person 1, every *second* person is eliminated — person 2 first, then 4, then 6, on around the shrinking ring — until one survivor remains.

Which seat survives? For $n = 10$ it is seat 5; for $n = 13$, seat 11. The survivors $1, 1, 3, 1, 3, 5, 7, 1, \ldots$ jump around with no obvious formula. **Is there a rule?**

Simulating the circle for large n is a bookkeeping nightmare. But watch what a single lap accomplishes.

THE MOVE

Halve, then relabel. After one trip around a circle of $2n$ people, all n *even* seats are gone and the count is poised to fall on person 1 again — a fresh Josephus circle of n people. The survivor of the small circle *is* the survivor of the big one; you only have to translate its label back. That translation is the entire recurrence, and it links size $2n$ (or $2n + 1$) directly to size n — a *halving* reduction rather than a one-step one.

WATCH THIS #4 The halving recurrence

Even case, $2n$ people. One lap eliminates $2, 4, \ldots, 2n$. The n survivors are the odds $1, 3, \ldots, 2n-1$, and the next to be counted is person 1 — exactly a new game of size n. If that game's survivor is seat $J(n)$, its label here is the $J(n)$-th odd number:

$$J(2n) = 2J(n) - 1.$$

Odd case, $2n + 1$ people. Eliminate $2, 4, \ldots, 2n$; then person $2n + 1$ is counted safe and person 1 falls next. The n survivors are $3, 5, \ldots, 2n + 1$, resuming at 3, so the reduced seat $J(n)$ maps to $2J(n) + 1$:

$$J(2n + 1) = 2J(n) + 1, \qquad J(1) = 1.$$

A closed form. Write $n = 2^m + \ell$ with $0 \le \ell < 2^m$ (so 2^m is the largest power of two not exceeding n). Both cases fuse into one clean rule, provable in a line by induction:

$$J(n) = 2\ell + 1.$$

For $n = 100 = 64 + 36$: $\ell = 36$, so $J(100) = 73$. ■

There is an even prettier way to read $J(n) = 2\ell + 1$. Write n in binary and slide its leading 1 to the *end*: that single cyclic shift *is* the answer. For $n = 100 = 1100100_2$, moving the front 1 to the back gives $1001001_2 = 73$. Chopping the top bit subtracts 2^m (leaving ℓ); re-appending it at the bottom doubles and adds one ($2\ell + 1$). *The survivor's seat is the starting number with its highest bit rotated to the bottom.*

Notice *why* the recurrence was the right instrument. Nothing in the scattered values $1, 1, 3, 1, 3, 5, 7, 1, \ldots$ suggests a formula, and blindly simulating the circle costs on the order of n^2 steps. Asking instead what a *whole lap* does to the problem's size turned an intractable-looking search into a two-line rule and then a one-line formula. *Whenever a process removes items in rounds, ask what one full round does to the size.* The reduction is almost always a recurrence.

SEEN BEFORE?

This is the "isolate the structural move, then recurse" idea of the Tower of Hanoi — but the reduction *halves* the problem, the divide-and-conquer shape behind recurrences like $T(n) = 2T(n/2) + \cdots$ that measure an algorithm's cost. And the binary-rotation closed form is the base-two thinking of **Chapter 9** slipping quietly into a counting puzzle.

Reading a recurrence's fingerprints

A recurrence is more than a way to compute. Its *shape* tells you what kind of answer to expect. A relation like $a_n = a_{n-1} + a_{n-2}$ (each term a sum of the two before) produces exponential growth at the golden ratio; a relation like $a_n = 2a_{n-1} + 1$ produces powers of two. Learning to read these fingerprints lets you guess a closed form and then confirm it.

For the Fibonacci-type tiling, the ratio of consecutive terms creeps toward a constant: $\frac{34}{21} \approx 1.619$, homing in on the *golden ratio* $\varphi = \frac{1+\sqrt{5}}{2} \approx 1.618$. That's not a coincidence — it's the recurrence's signature. Whenever consecutive terms settle into a fixed ratio, you're looking at exponential growth, and the ratio is the base. The next section makes this precise: the *characteristic equation* turns that limiting ratio into an exact closed form. For now, carry the instinct: *the recurrence predicts the growth rate before you ever find a formula.*

△ TRAPDOOR

A recurrence is useless without enough *starting values*. A two-term relation like $a_n = a_{n-1} + a_{n-2}$ needs *two* seeds (a_0 and a_1); with only one, the sequence isn't determined. And those seeds must match your problem's actual small cases — get them wrong and every later value is wrong, even with a perfect recurrence. Always pin down the base cases by hand, and double-check them against direct counts.

CURIOUS CARL • TRAPDOOR!

Carl writes down the correct rule $a_n = a_{n-1} + a_{n-2}$, pins the single seed $a_1 = 1$, and starts cranking: "one seed, one rule — the whole sequence is decided!"

The half-built sequence shrugs: *"To take a step you reach back* two *places — and you never told me where I stood before I started."* A two-term recurrence needs *two* seeds; with only a_1 fixed, a_2 is still free, and every later term drifts along with it.

From fingerprint to formula: the characteristic equation

The fingerprint told us *how fast* a sequence grows; now we extract *exactly* what it equals. The method is startlingly mechanical. For a two-term linear recurrence

$$a_n = A\,a_{n-1} + B\,a_{n-2},$$

guess that a pure geometric sequence $a_n = r^n$ might satisfy it, and see what r is forced to be. Substituting and dividing through by r^{n-2} collapses the entire recurrence into one quadratic:

$$r^n = Ar^{n-1} + Br^{n-2} \implies r^2 = Ar + B.$$

THE MOVE

The characteristic equation. For $a_n = Aa_{n-1} + Ba_{n-2}$, the quadratic $r^2 = Ar + B$ is the *characteristic equation*. If it has two distinct roots r_1, r_2, then *every* solution of the recurrence has the form

$$a_n = \alpha r_1^n + \beta r_2^n,$$

and the two constants α, β are pinned down by the two starting values. (Because the recurrence is *linear*, any blend of solutions is again a solution — which is why two geometric pieces suffice to capture them all.)

WATCH THIS #5 Binet's formula for the Fibonacci numbers

The setup. Fibonacci obeys $F_n = F_{n-1} + F_{n-2}$, so $A = B = 1$ and the characteristic equation is

$$r^2 = r + 1, \qquad \text{with roots} \qquad \varphi = \frac{1+\sqrt{5}}{2}, \quad \psi = \frac{1-\sqrt{5}}{2}.$$

These are the golden ratio and its conjugate — precisely the number the growth fingerprint pointed at.

Fit the seeds. Every solution is $F_n = \alpha\varphi^n + \beta\psi^n$. Impose $F_0 = 0$ and $F_1 = 1$:

$$\alpha + \beta = 0, \qquad \alpha\varphi + \beta\psi = 1.$$

The first gives $\beta = -\alpha$; substituting, $\alpha(\varphi - \psi) = 1$. Since $\varphi - \psi = \sqrt{5}$, we get $\alpha = 1/\sqrt{5}$ and $\beta = -1/\sqrt{5}$.

Binet's formula.

$$F_n = \frac{\varphi^n - \psi^n}{\sqrt{5}}.$$

A formula built entirely from $\sqrt{5}$ somehow returns a whole number at every step — check $n = 5$: $\varphi^5 \approx 11.09$, $\psi^5 \approx -0.09$, and $(11.09 + 0.09)/\sqrt{5} = 5 = F_5$. And because $|\psi| < 1$, the ψ^n term shrinks to nothing, so F_n is simply the nearest integer to $\varphi^n/\sqrt{5}$. ■

The growth fingerprint and the closed form are now one story: the base of the exponential *is* the dominant root of the characteristic equation, and the limiting ratio $F_{n+1}/F_n \to \varphi$ is just that root showing through.

△ TRAPDOOR

The two-geometric-pieces formula needs *distinct* roots. If the characteristic equation has a *re-*

peated root r, the two basic solutions are r^n and $n\,r^n$ — the second carries an extra factor of n. Writing $\alpha r^n + \beta r^n$ with the same r twice collapses into a single geometric term and cannot match two independent seeds.

Reversing the arrow: power sums

We have been driving one way down a street: a recurrence hands you a characteristic equation, whose roots hand you a closed form. Now drive it *backward*. Start with the roots of an equation and ask about their *power sums* $p_k = r^k + s^k + \cdots$ — and the recurrence comes roaring back, running the whole show.

Suppose r and s are the two roots of $x^2 - x - 1 = 0$ and you want $r^{100} + s^{100}$. The standard reflex is *Newton's sums*: grind out $p_1, p_2, p_3, \ldots$ one at a time from the symmetric functions of the roots — a bookkeeping slog that never ends, and hopeless when the exponent is 100. But the clean route is hiding inside the equation itself.

THE MOVE

Power sums obey the characteristic recurrence. If r, s are the roots of $x^2 = Ax + B$, then the power sums $p_k = r^k + s^k$ satisfy the *same* recurrence,

$$p_k = A\,p_{k-1} + B\,p_{k-2}, \qquad p_0 = 2, \quad p_1 = r + s = A.$$

The reason is one line: each root has $r^k = Ar^{k-1} + Br^{k-2}$ (multiply $r^2 = Ar + B$ by r^{k-2}); add the twin for s and the powers fuse into $p_k = Ap_{k-1} + Bp_{k-2}$. (More generally, for a degree-d polynomial whose roots each satisfy $r^d = c_1 r^{d-1} + c_2 r^{d-2} + \cdots + c_d$, the power sums $p_k = \sum_i r_i^k$ obey the same-shaped d-term recurrence $p_k = c_1 p_{k-1} + c_2 p_{k-2} + \cdots + c_d p_{k-d}$, its coefficients read straight off the polynomial. Writing the polynomial in this "$r^d = \cdots$" form fixes the signs; a polynomial given as $x^d + a_1 x^{d-1} + \cdots + a_d = 0$ has $c_i = -a_i$.) So drop Newton's sums and just *iterate a linear recurrence* — and since it's pure addition, you can reduce mod m at every step.

WATCH THIS — #6 The golden roots, to the hundredth power

Setup. Let r, s be the roots of $x^2 - x - 1 = 0$ — the golden ratio $\varphi \approx 1.618$ and its conjugate $\psi \approx -0.618$ (so $r + s = 1$, $rs = -1$). Their power sums obey $p_k = p_{k-1} + p_{k-2}$ with $p_0 = 2, p_1 = 1$: the *Lucas numbers* $2, 1, 3, 4, 7, 11, 18, \ldots$.

$r^{100} + s^{100}$ mod 11. Run the recurrence *mod* 11: $2, 1, 3, 4, 7, 0, 7, 7, 3, 10$, and then $2, 1, \ldots$ — it repeats with period 10. Since $100 \equiv 0 \pmod{10}$, $p_{100} \equiv p_0 = 2$. So $r^{100} + s^{100} \equiv \boxed{2} \pmod{11}$ — and we never wrote $\sqrt{5}$.

$\lfloor r^{200} \rfloor$ mod 7. Because $|\psi| < 1$, s^{200} is a *tiny positive* number in $(0, 1)$, so $r^{200} = p_{200} - s^{200}$ sits just below the integer p_{200}: hence $\lfloor r^{200} \rfloor = p_{200} - 1$. Reduce mod 7: the Lucas sequence mod 7 has period 16, and $200 \equiv 8 \pmod{16}$, where $p_8 = 47 \equiv 5$. Therefore $\lfloor r^{200} \rfloor \equiv 5 - 1 = \boxed{4} \pmod{7}$.

■

Trigger An expression $r^k + s^k$ (or $\sum_i r_i^k$) in the roots of a known polynomial — especially with a giant exponent, or a "mod m" or floor twist.

Do first Read the recurrence straight off the polynomial, seed with p_0 = (number of roots) and p_1 = (sum of roots), and iterate — reducing mod m as you go.

Trapdoor Getting the seeds wrong: $p_0 = 2$, *not* 1, and $p_1 = r + s$. And when a root has $|s| < 1$, remember s^k is the small correction that decides a floor.

Upgrade The same recurrence evaluates $\sum_i r_i^k$ for *any* polynomial and powers periodic "mod m" problems; the $|s| < 1$ correction is exactly why F_n is the nearest integer to $\varphi^n / \sqrt{5}$.

SEEN BEFORE?

This is the characteristic equation of this chapter *read backward*: before, the recurrence produced the roots; here, the roots reproduce the recurrence. The "iterate mod m until it cycles" step is the periodicity of **Chapters 9–10** — a linear recurrence mod m must repeat, since only finitely many pairs (p_{k-1}, p_k) exist — and the $|s| < 1$ floor trick is Binet's formula in work clothes.

The showpiece: Catalan, the recursive way

In Chapter 4 we met the Catalan numbers through a clever bijection. Here they appear again — by a completely different route — and seeing the *same sequence* fall out of a recurrence is one of the quiet joys of this subject.

WONDER THIS

You have $n+1$ numbers in a row that you want to multiply together, like $a \cdot b \cdot c \cdot d$. Multiplication only happens two-at-a-time, so you must insert parentheses to say what gets multiplied first: $(ab)(cd)$, or $a(b(cd))$, or $((ab)c)d$, and so on.

How many different ways can you fully parenthesize the product?

For $a \cdot b \cdot c$ there are 2 ways: $(ab)c$ and $a(bc)$. For four factors there are 5. Those numbers — $1, 2, 5, \ldots$ — are Catalan again. Let's see why, recursively.

WATCH THIS **#7 Splitting at the last multiplication**

Split on the outermost product. However you parenthesize $a_0 a_1 \cdots a_n$, there is exactly one *final* multiplication — the one performed last, joining a fully-parenthesized left chunk with a fully-parenthesized right chunk. Suppose that last multiply splits the row right after position i: the left chunk holds the first $i + 1$ factors, the right chunk the rest.

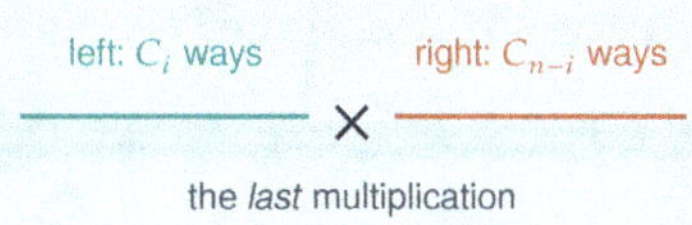

The recurrence. The left chunk can be parenthesized in C_i ways and the right chunk in C_{n-i}

ways, independently — so that split contributes $C_i \cdot C_{n-i}$. Summing over every place the final multiply could fall:

$$C_{n+1} = \sum_{i=0}^{n} C_i \, C_{n-i}, \qquad C_0 = 1.$$

The check. Build it up: $C_1 = C_0C_0 = 1$; $C_2 = C_0C_1 + C_1C_0 = 2$; $C_3 = 1{\cdot}2 + 1{\cdot}1 + 2{\cdot}1 = 5$; then $14, 42, 132, \ldots$ — the Catalan numbers exactly, matching the bijection count from Chapter 4. ■

This *convolution* recurrence — where each term is a sum of products of earlier terms — is the unmistakable fingerprint of a structure that splits into an independent left part and right part. Binary trees, triangulated polygons, balanced parentheses, our ticket-counter paths: every one of them obeys exactly this recurrence, because every one of them "splits in the middle" the same way. The bijections of Chapter 4 explained why these objects are *equinumerous*; the recurrence here explains the *mechanism* they all share. Two lenses, one truth. And that the answers agree is a small miracle you get to witness twice.

SEEN BEFORE?

This is the same Catalan sequence from **Chapter 4** (Bijections), now arising from a split-in-the-middle recurrence instead of a path-reflection. In **Chapter 8** (Generating Functions) we'll feed *this very recurrence* into a single equation $C(x) = 1 + x\,C(x)^2$ and solve for all the Catalan numbers at once — the third and most powerful view of the same objects.

TRY TO BREAK IT

A two-term recurrence needs two seeds — so try to starve one. Keep Fibonacci's rule $a_n = a_{n-1} + a_{n-2}$ but fix only $a_1 = 1$, leaving a_2 unnamed. Now grind out $a_3 = a_2 + 1$, $a_4 = 2a_2 + 1$, $a_5 = 3a_2 + 2$: every term is a *different* answer for every choice of a_2. Where does a single missing seed stop being a nuisance and start meaning the sequence simply doesn't exist yet?

Payoff: slicing the plane

Let's end with a recurrence you can almost see with your eyes.

WONDER THIS

Draw n straight lines across a sheet of paper, positioned so that every pair crosses and no three meet at a single point.

Into how many regions do the lines cut the plane? One line makes 2 regions, two lines make 4, three lines make...?

Add the lines one at a time and ask what the newest line contributes. When you draw the k-th line, it crosses each of the $k-1$ lines already present, at $k-1$ distinct points. Those crossing points chop the new line into k pieces (think: $k-1$ cuts make k segments). And here's the key: *each piece of the new line slices one existing region into two*, adding exactly one new region per piece. So the k-th line adds k regions.

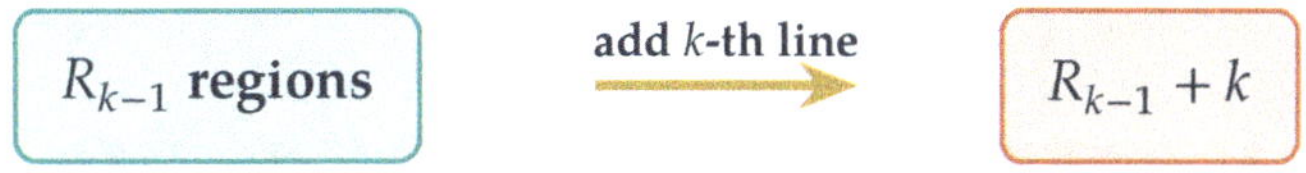

So $R_k = R_{k-1} + k$, starting from $R_0 = 1$ (the blank plane is one region). Unrolling, $R_n = 1 + (1 + 2 + \cdots + n) = 1 + \frac{n(n+1)}{2}$. Check it: $n = 3$ gives $1 + 6 = 7$ regions, exactly what you'd draw. **Each new line pays its own way, adding regions equal to its number** — and that triangular-number sum is the recurrence made visible. *The plane, it turns out, keeps remarkably honest books.*

What to try when you're stuck

When a problem is indexed by a size n and feels hard to attack all at once, set up a recurrence:

1. **Compute small cases by hand** ($n = 0, 1, 2, 3, 4$). Look up the sequence in your memory — Fibonacci? Powers of two? Catalan?
2. **Find the splitting feature.** What single decision (last column, first move, biggest element, final multiplication) breaks a size-n case into smaller ones?
3. **Make sure the cases are clean:** every large case lands in exactly one smaller case, with no overlaps and no omissions.
4. **Write the recurrence and pin the base cases** by direct count.
5. **Compute up, or guess a closed form** from the recurrence's growth fingerprint and verify it satisfies the relation.

THE MOVE

One-page recap. A recurrence defines each value from earlier ones by asking how a size-n problem reduces. Toolbox: **split on the last piece** (Fibonacci tilings), **split on the key move** (Hanoi, $H_n = 2H_{n-1} + 1 = 2^n - 1$), **halve and relabel** (Josephus, $J(n) = 2\ell + 1$), **convolution split** (Catalan, $C_{n+1} = \sum C_i C_{n-i}$), and **add one at a time** (lines cutting the plane, $R_n = 1 + \binom{n+1}{2}$). Read the recurrence's shape to predict growth; pin base cases by hand. When stuck, ask: *how does this shrink?*

THE NAME FOR IT

Recursion and recurrences

An equation defining each term from earlier ones is a *recurrence relation*; solving a problem by reducing it to a smaller copy of itself is *recursion*. The tiling counts here are the *Fibonacci numbers*, and the self-referential split that produced the Catalan count is a *convolution recurrence*.

WHERE IT LEADS Recurrences are *difference equations*, the discrete shadow of *differential equations*:

the characteristic-equation method here is exactly how linear ODEs are solved. They drive the *analysis of algorithms* (the Master Theorem, divide-and-conquer) and the spectral theory of linear operators.

CURIOUS CARL • EUREKA

Carl opens the biggest doll and finds a smaller one already holding the answer. "So I don't do the work — I make a tinier version of me do it and take the credit. Honestly? I can respect that."

SAME BEAST, NEW COSTUME

The same "shrink it, then build back up" beast, in costume across the book:

Fibonacci tilings → Tower of Hanoi → Josephus circle → Catalan convolution → Lines cutting the plane → Characteristic equation.

Each one refuses to attack size n head-on. It asks how the problem reduces to smaller copies of itself, and lets the small answers assemble the big one.

The practice ladder

THE PRACTICE LADDER *climb at your own pace*

LAUNCH *Warm the engine. Nudges below.*

1. You climb a staircase taking 1 or 2 steps at a time. In how many ways can you climb n stairs? Set up the recurrence and identify the sequence.
2. A sequence has $a_1 = 3$ and $a_n = 2a_{n-1}$. Find a_6 and a closed form.
3. Prove by induction that $1 + 3 + 5 + \cdots + (2n-1) = n^2$ for all $n \geq 1$. State the base case and the inductive step explicitly.

CORE BUILDER *Master the central moves.*

1. How many length-n strings of 0s and 1s have *no two adjacent* 1s? Find the recurrence and connect it to Fibonacci.
2. Solve $a_n = 2a_{n-1} + 3$ with $a_0 = 0$ by computing terms and spotting the pattern; then verify your closed form.
3. Count the regions formed inside a circle by n chords in "general position" (every two cross inside, no three concurrent). Build the recurrence.
4. Prove by induction that $n < 2^n$ for every integer $n \geq 1$.
5. Solve $a_n = 5a_{n-1} - 6a_{n-2}$ with $a_0 = 0$, $a_1 = 1$ using the characteristic equation, and give a closed form.

STRETCH *Transfer to unfamiliar ground.*

1. A *derangement* is a permutation leaving nothing in its original spot. Prove $D_n = (n-1)(D_{n-1} + D_{n-2})$ by splitting on where element 1 lands.

2. In how many ways can a convex $(n+2)$-gon be cut into triangles by non-crossing diagonals? Set up the Catalan recurrence by splitting on the triangle attached to one fixed edge.

3. The *Pell numbers* satisfy $P_n = 2P_{n-1} + P_{n-2}$ with $P_0 = 0$, $P_1 = 1$. Use the characteristic equation to derive a Binet-style closed form for P_n.

4. Let r, s be the roots of $x^2 - x - 1 = 0$. Show the power sums $p_k = r^k + s^k$ satisfy $p_k = p_{k-1} + p_{k-2}$, and use it to compute $r^5 + s^5$ (no $\sqrt{5}$ needed).

THE FLASH *No clock — each cracks open in one flash of insight, once you find the angle. Nudges at the back.*

1. How many tilings does a 2×10 strip have?

2. Solve the Tower of Hanoi for $n = 8$: how many moves?

3. Let r, s be the roots of $x^2 - 3x + 1 = 0$. Using the power-sum recurrence, find $r^4 + s^4$.

PROOF LAB *Write it clean, in full sentences.*

1. Prove that the $2 \times n$ domino tiling count equals the Fibonacci number F_{n+1}, with a careful argument that the "last column" split is exhaustive and non-overlapping.

2. Prove $\sum_{i=0}^{n} C_i C_{n-i} = C_{n+1}$ counts full parenthesizations of $n+2$ factors, and explain why the same recurrence governs binary trees.

3. Prove, by strong induction, that every amount of postage of 12 cents or more can be made using only 4-cent and 5-cent stamps.

4. Let r, s be the roots of $x^2 - x - 1 = 0$. Prove that $p_n = r^n + s^n$ is an integer for every $n \geq 0$ by showing it satisfies an integer-seeded recurrence; identify the sequence.

5. *The halving recurrence.* For the standard every-second Josephus problem, prove $J(2n) = 2J(n) - 1$ and $J(2n+1) = 2J(n) + 1$ with $J(1) = 1$, and deduce that $J(n) = 2\ell + 1$ when $n = 2^m + \ell$ with $0 \leq \ell < 2^m$. Find the surviving seat for $n = 1000$.

6. *Josephus, every third (a variation).* Now eliminate every *third* person. Numbering the seats $0, 1, \ldots, n-1$, the survivor satisfies $J(n) = (J(n-1) + 3) \bmod n$ with $J(1) = 0$. Explain why the index shifts by 3 and reduces modulo n, compute $J(n)$ for $n = 1, \ldots, 7$, and state the surviving seat (counted from 1) for $n = 7$.

REFLECTION *No nudge. Write a few sentences.*

1. For one problem above, what feature did you split on — or what made induction the right tool? Describe how you confirmed the cases were exhaustive and non-overlapping, or how you chose your base case(s).

NUDGES

Launch. (1) $a_n = a_{n-1} + a_{n-2}$, the Fibonacci numbers (with $a_1 = 1, a_2 = 2$). (2) $a_6 = 3 \cdot 2^5 = 96$; closed form $a_n = 3 \cdot 2^{n-1}$. (3) Base: $1 = 1^2$. Step: adding the next odd number, $k^2 + (2k+1) = (k+1)^2$. **Core.** (4) $f_n = f_{n-1} + f_{n-2}$ (append 0, or append 01); Fibonacci again. (5) Terms $0, 3, 9, 21, 45, \ldots$; closed form $a_n = 3(2^n - 1)$. (6) $R_n = R_{n-1} + n$ with $R_0 = 1$, giving $1 + \binom{n+1}{2}$. (7) Base $1 < 2$; step: $k < 2^k \Rightarrow k + 1 \le 2^k + 1 \le 2^k + 2^k = 2^{k+1}$. (8) Characteristic $r^2 - 5r + 6 = (r-2)(r-3)$; $a_n = \alpha 2^n + \beta 3^n$, and the seeds give $\alpha = -1$, $\beta = 1$, so $a_n = 3^n - 2^n$. **Stretch.** (9) Element 1 goes to some spot j; either j returns to 1 (gives D_{n-2}) or not (gives D_{n-1}), and j has $n-1$ choices. (10) Fix an edge; the triangle on it splits the polygon into two smaller polygons, giving the convolution. (11) $r^2 = 2r + 1$, roots $1 \pm \sqrt{2}$; fitting the seeds, $P_n = \dfrac{(1+\sqrt{2})^n - (1-\sqrt{2})^n}{2\sqrt{2}}$. (12) Each root gives $r^k = r^{k-1} + r^{k-2}$, so $p_k = p_{k-1} + p_{k-2}$; with $p_0 = 2, p_1 = 1$ (the Lucas numbers), $p_5 = 11$. **The Flash.** (13) $F_{11} = 89$. (14) $2^8 - 1 = 255$. (15) $x^2 = 3x - 1$ gives $p_k = 3p_{k-1} - p_{k-2}$, with $p_0 = 2, p_1 = 3$: the terms run $2, 3, 7, 18, 47$, so $r^4 + s^4 = 47$. **Proof.** (16) Let T_n be the number of tilings. The last column's top cell is covered either by a vertical domino filling that column — leaving a $2 \times (n-1)$ strip — or by a horizontal domino, which forces a second horizontal beneath it and leaves a $2 \times (n-2)$ strip. The two cases are exhaustive and disjoint, so $T_n = T_{n-1} + T_{n-2}$; with $T_1 = 1, T_2 = 2$ this is the Fibonacci recurrence, giving $T_n = F_{n+1}$. (17) Write a_m for the number of full parenthesizations of m factors. The single outermost multiplication splits the factors into a left block of k and a right block of $m-k$, each independently parenthesized, so $a_m = \sum_{k=1}^{m-1} a_k a_{m-k}$; with $a_m = C_{m-1}$ and $m = n+2$ this is exactly $C_{n+1} = \sum_{i=0}^{n} C_i C_{n-i}$. A binary tree's root likewise splits into independent left and right subtrees, giving the identical convolution. (18) Show every $n \ge 12$ is a non-negative combination of 4s and 5s. The base cases $12 = 4{+}4{+}4$, $13 = 4{+}4{+}5$, $14 = 4{+}5{+}5$, $15 = 5{+}5{+}5$ cover four consecutive values; for the step, any $n \ge 16$ has $n - 4 \ge 12$, which is makeable by the strong hypothesis, so adding one 4-cent stamp makes n. The four bases guarantee $n - 4$ has already been settled, so the induction closes. (19) From $r^2 = r + 1$ (and likewise for s), multiply by r^{n-2} to get $r^n = r^{n-1} + r^{n-2}$; summing the r and s versions gives $p_n = p_{n-1} + p_{n-2}$. The seeds $p_0 = 2$ and $p_1 = r + s = 1$ are integers, so by induction every p_n is an integer — the Lucas numbers $2, 1, 3, 4, 7, 11, \ldots$. (20) Number the seats $1, \ldots, 2n$. The first lap eliminates every even seat, leaving the n odd seats with seat 1 next to be skipped, so the rest is a size-n game whose winner $J(n)$ sits at the $J(n)$-th odd seat, $2J(n) - 1$; thus $J(2n) = 2J(n) - 1$. For $2n + 1$ seats the extra person makes seat 1 fall too, shifting the survivor to $2J(n) + 1$. Iterating from $J(1) = 1$ peels off the leading binary bit, giving $J(2^m + \ell) = 2\ell + 1$; for $n = 1000 = 512 + 488$ this is $2 \cdot 488 + 1 = 977$. (21) After one elimination, $n - 1$ people remain and counting resumes 3 seats past the start, so if the reduced circle's survivor sits at index $J(n-1)$ it lies 3 further along the full circle, wrapping around: $J(n) = (J(n-1) + 3) \bmod n$ with $J(1) = 0$. Iterating gives $J(1), \ldots, J(7) = 0, 1, 1, 0, 3, 0, 3$, so $J(7) = 3$ in 0-indexing — the survivor is seat 4 counting from 1.

CHAPTER SIX

6

The Great Cancellation

The best way to add a thousand things is to make nine hundred ninety-eight of them cancel.

Telescoping: make the middle disappear.

Here is a sum that looks like it should get messier as it grows. It does exactly the opposite.

WONDER THIS

Add up the fractions

$$\frac{1}{1\cdot 2}+\frac{1}{2\cdot 3}+\frac{1}{3\cdot 4}+\cdots+\frac{1}{n(n+1)}.$$

What does this equal? Try a few: the first term is $\frac{1}{2}$, the first two sum to $\frac{2}{3}$, the first three to $\frac{3}{4}$...

CURIOUS CARL • FIRST GUESS

Carl: "A hundred fractions — one common denominator, one enormous pile, and I grind the whole thing down to a single sum."

The catch: that's a hundred chances to slip, Carl — but rewrite each term $\frac{1}{k(k+1)}$ as a *difference*, and watch the whole middle cancel itself away, leaving only the two ends.

The partial sums are $\frac{1}{2}, \frac{2}{3}, \frac{3}{4}, \frac{4}{5}, \ldots$. The pattern is screaming at you. It seems the sum of the first n terms is just $\frac{n}{n+1}$. But *why*? Adding fractions usually makes them uglier, not cleaner. The secret is to look harder at a single term.

Each fraction splits in two. Watch:

$$\frac{1}{k(k+1)} = \frac{1}{k} - \frac{1}{k+1}.$$

(Check: $\frac{1}{k} - \frac{1}{k+1} = \frac{(k+1)-k}{k(k+1)} = \frac{1}{k(k+1)}$.) Now rewrite the whole sum with every term cracked open:

$$\left(\frac{1}{1} - \frac{1}{2}\right) + \left(\frac{1}{2} - \frac{1}{3}\right) + \left(\frac{1}{3} - \frac{1}{4}\right) + \cdots + \left(\frac{1}{n} - \frac{1}{n+1}\right).$$

Look what happens in the middle: the $-\frac{1}{2}$ from the first bracket annihilates the $+\frac{1}{2}$ from the second; the $-\frac{1}{3}$ kills the $+\frac{1}{3}$; every interior term cancels its neighbor. The entire sum collapses like a folding spyglass, leaving only the very first piece and the very last:

$$\frac{1}{1} - \frac{1}{n+1} = \frac{n}{n+1}.$$

Almost everything cancels, and only the two endpoints survive. That collapse is called **telescoping**, after the old brass telescopes that slide shut into a fraction of their length.

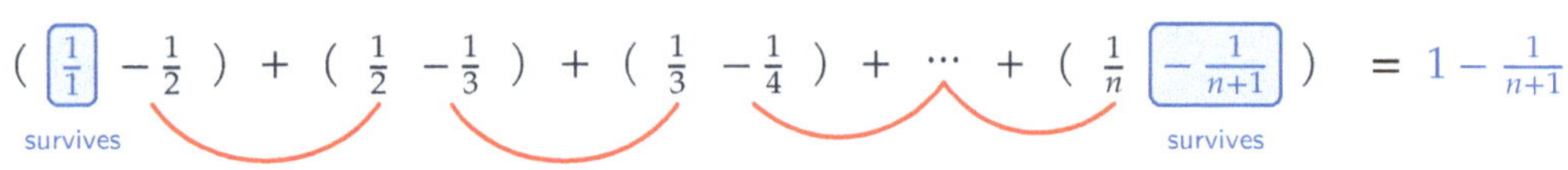

Each term splits into a difference; neighbouring pieces cancel, and only the two endpoints survive.

THE TURN *from grind to insight*

The obvious way to face a sum like this is to add it the honest way — find a common denominator, combine the first two fractions, fold in the third, and keep grinding term after term, hoping a closed form eventually shows itself. For a thousand terms that is a thousand chances to slip, and the pattern only emerges after enormous labor. **The creative turn** is to stop adding and instead *rewrite each single term* as a difference, $\frac{1}{k(k+1)} = \frac{1}{k} - \frac{1}{k+1}$. Once every term is a gap between two consecutive values, neighbors annihilate in a chain and only the very first and very last pieces survive. We never added the sum — we engineered it to add itself to almost nothing. When a long sum has a suspiciously clean answer, don't compute it; find the difference hiding inside each term.

THE MOVE

A sum **telescopes** when each term can be written as a *difference* $f(k)-f(k+1)$ (or $f(k+1)-f(k)$). Then

$$\sum_{k=1}^{n}(f(k)-f(k+1))=f(1)-f(n+1),$$

because every interior value is created once and destroyed once. The point is *recognizing each term as a difference*: finding the right f. Once you see it, a sum of a thousand terms has a two-term answer.

This is the laziest kind of brilliance: you don't add the terms, you arrange for them to add themselves to zero. The art is never the arithmetic — it's spotting that a complicated-looking term is secretly the gap between two consecutive values of something simpler.

Manufacturing the difference

In the opener, the splitting $\frac{1}{k(k+1)}=\frac{1}{k}-\frac{1}{k+1}$ appeared like magic. It isn't magic — it's *partial fractions*, a routine technique for breaking a fraction into a difference. When the denominator factors, you can almost always engineer a telescoping difference.

WATCH THIS **#1 A wider gap**

Claim. $\displaystyle\sum_{k=1}^{n}\frac{1}{k(k+2)}=\frac{1}{2}\left(\frac{3}{2}-\frac{1}{n+1}-\frac{1}{n+2}\right).$

Split the term. The factors are now two apart, so the difference spans two steps:

$$\frac{1}{k(k+2)}=\frac{1}{2}\left(\frac{1}{k}-\frac{1}{k+2}\right).$$

(The $\frac{1}{2}$ is forced: $\frac{1}{k}-\frac{1}{k+2}=\frac{2}{k(k+2)}$, so halve it.)

Telescope with a stride of 2. Now the cancellation skips a step — each $-\frac{1}{k+2}$ cancels the $+\frac{1}{k+2}$ that appears *two* terms later. So instead of one surviving term at each end, *two* survive at each end:

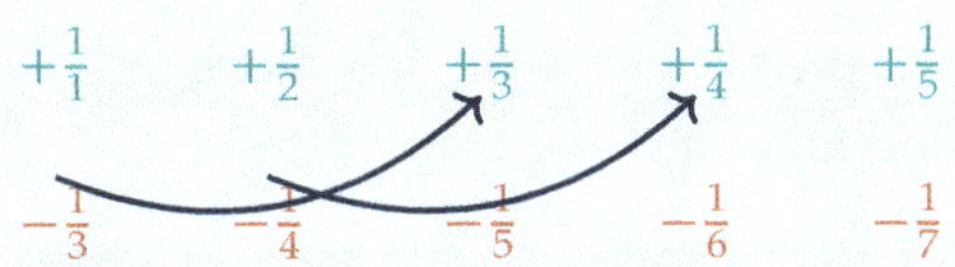

$+\frac{1}{1}$, $+\frac{1}{2}$ **survive at the front**

The survivors are the two opening positives $\frac{1}{1}+\frac{1}{2}$ and the two closing negatives $-\frac{1}{n+1}-\frac{1}{n+2}$, all scaled by $\frac{1}{2}$:

$$\frac{1}{2}\left(1+\frac{1}{2}-\frac{1}{n+1}-\frac{1}{n+2}\right)=\frac{1}{2}\left(\frac{3}{2}-\frac{1}{n+1}-\frac{1}{n+2}\right).$$

For $n=5$ this gives $\frac{25}{42}$, which you can check by brute force. ∎

Trigger A sum of fractions whose denominators factor, or any term that looks like it could be a gap between two simpler values.

Do first Use partial fractions to write each term as $c\,(f(k)-f(k+d))$, then telescope; a stride of d leaves d survivors on each end.

Trapdoor Losing track of how many endpoint terms survive — a stride-2 difference does *not* collapse to a single term. Write out the first and last few explicitly.

Upgrade Any factored denominator yields a partial-fraction difference; the wider the factor gap, the more boundary terms remain.

The lesson generalizes cleanly: with a stride of d, the front contributes d surviving terms and so does the back. The collapse is just as total. It simply leaves a slightly thicker rim. Always write out the first three and last three terms before declaring what survives; the cancellation is reliable, but the *bookkeeping* at the edges is where people slip.

△ **TRAPDOOR**

The single most common telescoping error is miscounting the survivors. "Everything cancels except the ends" is true, but *how many* terms live at each end depends on the stride. A difference $f(k)-f(k+1)$ leaves one term per side; $f(k)-f(k+2)$ leaves two. When in doubt, expand the first and last few brackets by hand and see exactly what fails to find a partner.

CURIOUS CARL • TRAPDOOR!

Carl splits every term into a difference, watches the interior vanish, and declares the whole sum equal to 0: "It all cancels — there's nothing left!"

The two survivors wave from the edges: "*Everything* inside *paired off — but we, the very first term and the very last, never found a partner.*" Telescoping empties the middle, not the ends; a stride-1 collapse always leaves $f(1)-f(n+1)$ standing.

Telescoping in disguise: products and more

Telescoping isn't only for sums. The same "each piece cancels the next" collapse works for *products*, where instead of differences canceling, factors divide out.

WATCH THIS **#2 A product that folds shut**

Claim. $\displaystyle\prod_{k=2}^{n}\left(1-\frac{1}{k^2}\right)=\frac{n+1}{2n}.$

Factor each piece. The key is that $1-\frac{1}{k^2}$ is a difference of squares:

$$1-\frac{1}{k^2}=\frac{k^2-1}{k^2}=\frac{(k-1)(k+1)}{k\cdot k}=\frac{k-1}{k}\cdot\frac{k+1}{k}.$$

So each factor is itself a little product of two ratios. Write out the whole thing for $k = 2, 3, 4, \dots, n$:

$$\underbrace{\frac{1}{2}\cdot\frac{3}{2}}_{k=2}\cdot\underbrace{\frac{2}{3}\cdot\frac{4}{3}}_{k=3}\cdot\underbrace{\frac{3}{4}\cdot\frac{5}{4}}_{k=4}\cdots\underbrace{\frac{n-1}{n}\cdot\frac{n+1}{n}}_{k=n}.$$

Watch two telescopes at once. Separate the "$\frac{k-1}{k}$" factors from the "$\frac{k+1}{k}$" factors. The first group, $\frac{1}{2}\cdot\frac{2}{3}\cdot\frac{3}{4}\cdots\frac{n-1}{n}$, telescopes to $\frac{1}{n}$ (numerators cancel the next denominators). The second group, $\frac{3}{2}\cdot\frac{4}{3}\cdot\frac{5}{4}\cdots\frac{n+1}{n}$, telescopes to $\frac{n+1}{2}$. Multiply the two survivors:

$$\frac{1}{n}\cdot\frac{n+1}{2}=\frac{n+1}{2n}.$$

For $n = 5$: $\frac{6}{10} = \frac{3}{5}$. ■

This is the showpiece, and it's worth pausing on what made it work. A *product* of $n-1$ messy factors collapsed to a single clean ratio, because each factor, once split by difference-of-squares, handed a piece to its left neighbor and a piece to its right. Telescoping is fundamentally about *adjacency*: terms that talk to their neighbors and cancel. Sums telescope through subtraction; products telescope through division; and as you're about to see, even angles can telescope.

TRY TO BREAK IT

Telescoping lives or dies on *consecutive* differences. So aim at something that refuses to line up: try to collapse $\frac{1}{1\cdot 3}+\frac{1}{2\cdot 4}+\frac{1}{3\cdot 5}+\cdots+\frac{1}{k(k+2)}$ as if it were a stride-1 chain. Split each term $\frac{1}{k(k+2)}=\frac{1}{2}\left(\frac{1}{k}-\frac{1}{k+2}\right)$ and write out five terms: which negatives fail to meet the positive two steps ahead, and why does the collapse leave *two* survivors at each end instead of one? Push further — what happens if a denominator like $\frac{1}{k(k+3)}$ jumps by three?

Powering through powers: $\sum k^2$ and $\sum k^3$

Here is the cancellation idea at its most useful. Everyone memorizes $1^2+2^2+\cdots+n^2=\frac{n(n+1)(2n+1)}{6}$, but almost nobody remembers *where it comes from*. Telescoping hands it to you. And the method generalizes to any power.

WONDER THIS

You want a closed form for $1^2+2^2+\cdots+n^2$. You could try to guess a cubic and fit coefficients, but that feels like cheating.

Is there a way to *derive* it — one that also tells you how to get $\sum k^3$, $\sum k^4$, and beyond?

WATCH THIS 	**#3 The sum of squares falls out of a cube**

Start from a telescoping difference — of cubes. Consider $(k+1)^3 - k^3$. Summed from $k = 1$ to n, it telescopes instantly to $(n+1)^3 - 1$. But expand the same difference algebraically:

$$(k+1)^3 - k^3 = 3k^2 + 3k + 1.$$

So the *same* telescoping sum equals $\sum_{k=1}^{n}(3k^2 + 3k + 1)$. Setting the two evaluations equal:

$$3\sum_{k=1}^{n} k^2 + 3\sum_{k=1}^{n} k + n = (n+1)^3 - 1.$$

Everything here is known except $\sum k^2$. We already have $\sum k = \frac{n(n+1)}{2}$, so substitute and solve:

$$3\sum k^2 = (n+1)^3 - 1 - n - \tfrac{3n(n+1)}{2} \implies \sum_{k=1}^{n} k^2 = \frac{n(n+1)(2n+1)}{6}.$$

(Check $n = 5$: $\frac{5\cdot 6\cdot 11}{6} = 55 = 1 + 4 + 9 + 16 + 25$.) ■

The engine is reusable. To get $\sum k^3$, telescope one power higher: $(k+1)^4 - k^4 = 4k^3 + 6k^2 + 4k + 1$ sums to $(n+1)^4 - 1$, and now every piece is known *except* $\sum k^3$, so solving gives

$$\sum_{k=1}^{n} k^3 = \frac{n^2(n+1)^2}{4} = \left(\frac{n(n+1)}{2}\right)^2 = \Big(\sum_{k=1}^{n} k\Big)^2,$$

the lovely fact that the sum of the first n cubes is the *square of the sum* of the first n integers. Each new power leans on all the lower ones, a perfect cancellation staircase: telescope $(k+1)^{p+1} - k^{p+1}$, and the only unknown is $\sum k^p$.

WATCH THIS — #4 Why this becomes the integral

Look at the leading behavior. The exact answer $\sum_{k=1}^{n} k^2 = \frac{n(n+1)(2n+1)}{6}$ has leading term $\frac{n^3}{3}$. That fraction $\frac{1}{3}$ is not a coincidence. It is exactly $\int_0^1 x^2\,dx = \frac{1}{3}$, scaled up.
Here is the bridge. Slice $[0, n]$ into n unit-width rectangles, the k-th of height k^2. Their total area is $\sum k^2$, and it approximates the area under $y = x^2$:

$$\sum_{k=1}^{n} k^2 \approx \int_0^n x^2\,dx = \frac{n^3}{3}.$$

The telescoping identity $(k+1)^3 - k^3 = 3k^2 + 3k + 1$ is the *discrete twin* of the derivative $\frac{d}{dx}x^3 = 3x^2$: differencing x^3 produces $3x^2$ plus lower-order "correction" terms $(3k+1)$ that vanish in the limit. Summing a difference back up (telescoping) is the discrete Fundamental Theorem of Calculus; letting the slices shrink turns the sum into the integral and the corrections disappear, leaving the clean $\frac{n^3}{3}$. ■

This is the deepest payoff of the whole chapter: the trick of writing a term as a difference $f(k+1) - f(k)$ *is* differentiation in disguise, and telescoping it back *is* integration. The power-sum formulas are what you get before taking the limit; the integral $\int x^p\,dx = \frac{x^{p+1}}{p+1}$ is what survives after. Newton and Leibniz built calculus by pushing exactly this idea: from finite cancellation to the continuous limit.

The payoff(s): factorials and even angles

Two quick, satisfying collapses to close: one algebraic, one positively sneaky.

WONDER THIS

First: add up $1 \cdot 1! + 2 \cdot 2! + 3 \cdot 3! + \cdots + n \cdot n!$.

Second: add up the angles $\arctan \frac{1}{3} + \arctan \frac{1}{7} + \arctan \frac{1}{13} + \cdots$, where the denominators are $k^2 + k + 1$.

Both look hopeless. Both telescope.

For the factorials, the trick is to spot that $k \cdot k!$ is a difference of factorials in disguise. Since $(k+1)! = (k+1) \cdot k! = k \cdot k! + k!$, we get $k \cdot k! = (k+1)! - k!$ — a perfect telescoping difference! So

$$\sum_{k=1}^{n} k \cdot k! = \sum_{k=1}^{n} ((k+1)! - k!) = (n+1)! - 1!,$$

everything in between canceling. For $n = 6$ that's $7! - 1 = 5039$. **The whole sum is just one factorial, minus one.**

The angle sum is the sneaky one, and it shows how far the idea reaches. There's a classic identity, $\arctan a - \arctan b = \arctan \frac{a-b}{1+ab}$. Set $a = k+1$ and $b = k$: then $\frac{a-b}{1+ab} = \frac{1}{1+k(k+1)} = \frac{1}{k^2+k+1}$ — exactly our terms! (Here $a = k+1 > b = k > 0$, so the difference $\arctan(k+1) - \arctan(k)$ lies strictly between 0 and $\frac{\pi}{2}$, and the principal-value arctangent formula applies with no branch correction.) So each angle is a difference $\arctan(k+1) - \arctan(k)$, and the sum telescopes to $\arctan(n+1) - \arctan(1)$. *Even arctangents, it turns out, fold up like a spyglass.* The terms were differences all along; we just had to recognize the disguise.

Put on finite-difference glasses. Every example in this chapter is one idea wearing costumes. Write the *forward difference* of a function as $\Delta f(k) = f(k+1) - f(k)$. Then a telescoping sum is *exactly*

$$\sum_{k=1}^{n} \Delta f(k) = f(n+1) - f(1),$$

the interior canceling by the very definition of Δ. This is the fundamental theorem of calculus *before* limits: the difference Δ plays the role of the derivative, the sum plays the role of the integral, and "a sum of differences collapses to its endpoints" is the discrete twin of "the integral of a derivative is a difference of endpoint values." Seen this way, every telescoping problem is a single question (*find the f whose difference is your summand*) and "make the middle cancel" is just the discrete hunt for an antiderivative.

SEEN BEFORE?

Telescoping is the discrete cousin of the *Fundamental Theorem of Calculus*: a sum of differences collapses to its endpoints, just as an integral of a derivative collapses to boundary values. The "write each term as a difference" instinct also powers summation by parts, and it pairs beautifully with the recurrences of **Chapter 5**. Many recurrences are, once translated, telescoping in disguise. You'll see the same endpoint-survival idea in **Chapter 9**'s modular sums.

What to try when you're stuck

When you face a sum or product with many terms and a suspiciously clean-looking answer, suspect telescoping:

1. **Compute partial sums** for $n = 1, 2, 3$. A clean pattern (like $\frac{n}{n+1}$) is a telescoping fingerprint.
2. **Try to write each term as a difference** $f(k) - f(k+d)$. For fractions, use partial fractions; for factorials, look for $(k+1)! - k!$ patterns; for products, factor each piece.
3. **Write out the first three and last three terms** expanded, and see what cancels.
4. **Count the survivors:** a stride-d difference leaves d terms at each end.
5. **Add (or multiply) the survivors** — that's your closed form.

THE MOVE

One-page recap. A sum telescopes when each term is a difference $f(k) - f(k+d)$; then everything interior cancels and only d boundary terms survive at each end. Toolbox: partial fractions ($\frac{1}{k(k+1)} = \frac{1}{k} - \frac{1}{k+1}$), difference of squares for products ($1 - \frac{1}{k^2}$), factorial differences ($k \cdot k! = (k+1)! - k!$), the arctan subtraction formula, and power sums by telescoping $(k+1)^{p+1} - k^{p+1}$ (giving $\sum k^2$, $\sum k^3$, and the bridge to $\int x^p\,dx$). When a long sum has a short answer, find the difference hiding in each term.

THE NAME FOR IT

Telescoping sums

A sum whose interior terms cancel in pairs, leaving only the endpoints, is a *telescoping sum*. The technique of splitting a fraction into a difference of simpler ones is *partial fractions*, and the discrete cancellation mirrors the *Fundamental Theorem of Calculus*.

WHERE IT LEADS Telescoping is the discrete *Fundamental Theorem of Calculus*: a sum of differences collapses to its endpoints just as $\int_a^b F'(x)\,dx = F(b) - F(a)$. This duality between summation and differencing is formalized as the *calculus of finite differences* and *Abel summation*, central to numerical analysis.

CURIOUS CARL • EUREKA

Carl: "I brought a hundred fractions to a fight — ninety-eight of them cancelled each other out and went home, and I won by barely showing up."

SAME BEAST, NEW COSTUME

The same "middle cancels, ends survive" beast, in costume across the book:

Fraction sums → Product telescoping → Power sums → Summation by parts → The coupon-collector harmonic sum.

Each writes its terms as gaps between consecutive values of something simpler, so neighbors annihilate in a chain and only the boundary remains.

The practice ladder

THE PRACTICE LADDER *climb at your own pace*

LAUNCH *Warm the engine. Nudges below.*

1. Evaluate $\sum_{k=1}^{n}(k^2-(k-1)^2)$ by telescoping, and explain why the answer is just n^2.
2. Find $\sum_{k=1}^{20}\frac{1}{k(k+1)}$ using the opener's collapse.

CORE BUILDER *Master the central moves.*

1. Evaluate $\sum_{k=1}^{n}\frac{1}{k(k+1)(k+2)}$ by partial fractions. (Hint: it's $\frac{1}{2}(\frac{1}{k(k+1)}-\frac{1}{(k+1)(k+2)})$.)
2. Find a closed form for $\sum_{k=1}^{n}\frac{k}{(k+1)!}$ by writing each term as a difference of $\frac{1}{k!}$ values.
3. Evaluate $\prod_{k=2}^{n}\frac{k^3-1}{k^3+1}$ by factoring numerator and denominator and watching the collapse.

STRETCH *Transfer to unfamiliar ground.*

1. Derive $\sum_{k=1}^{n}k^2=\frac{n(n+1)(2n+1)}{6}$ by telescoping $k^3-(k-1)^3$ and solving for the square-sum.
2. Derive $\sum_{k=1}^{n}k^3=\left(\frac{n(n+1)}{2}\right)^2$ by telescoping $(k+1)^4-k^4$ and substituting the known lower power sums.
3. Evaluate $\sum_{k=1}^{n}\frac{k}{2^k}$ by finding a function $g(k)$ with $\frac{k}{2^k}=g(k)-g(k+1)$.

THE FLASH *No clock — each cracks open in one flash of insight, once you find the angle. Nudges at the back.*

1. Compute $\sum_{k=1}^{99}\frac{1}{\sqrt{k}+\sqrt{k+1}}$. (Rationalize each term first.)
2. Compute $\sum_{k=1}^{n}k\cdot k!$ in closed form.

PROOF LAB *Write it clean, in full sentences.*

1. Prove the telescoping identity $\sum_{k=1}^{n}(f(k)-f(k+1))=f(1)-f(n+1)$ by induction, and explain why it is the discrete analogue of the Fundamental Theorem of Calculus.

2. Prove $\sum_{k=1}^{n}\arctan\frac{1}{k^2+k+1}=\arctan(n+1)-\arctan 1$, justifying the use of the arctangent subtraction formula at each step.

3. Using $\sum_{k=1}^{n}k^2=\frac{n(n+1)(2n+1)}{6}$, show that $\frac{1}{n^3}\sum_{k=1}^{n}k^2\to\frac{1}{3}$ as $n\to\infty$, and explain why this limit equals $\int_0^1 x^2\,dx$. (This is the telescoping-to-integral bridge.)

REFLECTION *No nudge. Write a few sentences.*

1. For one sum above, describe how you found the hidden difference. What told you the term was secretly $f(k)-f(k+d)$?

NUDGES

Launch. (1) Telescopes to $n^2-0^2=n^2$. (2) $1-\frac{1}{21}=\frac{20}{21}$. **Core.** (3) Survivors give $\frac{1}{2}(\frac{1}{2}-\frac{1}{(n+1)(n+2)})$. (4) $\frac{k}{(k+1)!}=\frac{1}{k!}-\frac{1}{(k+1)!}$; sum $=1-\frac{1}{(n+1)!}$. (5) Factor $k^3\pm1$ via sum/difference of cubes; the product collapses to $\frac{2(n^2+n+1)}{3n(n+1)}$. **Stretch.** (6) $\sum(k^3-(k-1)^3)=n^3$ and the left side is $3\sum k^2-3\sum k+n$; solve. (7) $(k+1)^4-k^4=4k^3+6k^2+4k+1$ sums to $(n+1)^4-1$; substitute $\sum k^2,\sum k$ and solve for $\sum k^3=(\frac{n(n+1)}{2})^2$. (8) Take $g(k)=\frac{k+1}{2^{k-1}}$ (or similar); sum $=2-\frac{n+2}{2^n}$. **The Flash.** (9) Each term rationalizes to $\sqrt{k+1}-\sqrt{k}$; sum $=\sqrt{100}-1=9$. (10) $(n+1)!-1$. **Proof.** (11) Induction on n. Base case $n=1$: both sides equal $f(1)-f(2)$. Inductive step: adding the $(n+1)$st term $f(n+1)-f(n+2)$ to the assumed sum $f(1)-f(n+1)$ makes the $f(n+1)$ terms cancel, leaving $f(1)-f(n+2)$. It is the discrete Fundamental Theorem of Calculus because each summand is a difference Δf and the sum keeps only the endpoints, exactly as $\int_a^b f'=f(b)-f(a)$. (12) Since $\frac{1}{k^2+k+1}=\frac{(k+1)-k}{1+(k+1)k}$, the subtraction formula $\arctan a-\arctan b=\arctan\frac{a-b}{1+ab}$ with $a=k+1$, $b=k$ turns each term into $\arctan(k+1)-\arctan k$ (valid since $ab=k(k+1)\geq 0$ keeps the argument in principal range). Summing telescopes to $\arctan(n+1)-\arctan 1$. (13) $\frac{1}{n^3}\cdot\frac{n(n+1)(2n+1)}{6}=\frac{(n+1)(2n+1)}{6n^2}=\frac{2n^2+3n+1}{6n^2}\to\frac{2}{6}=\frac{1}{3}$. This equals $\int_0^1 x^2\,dx$ because $\frac{1}{n}\sum_{k=1}^{n}(k/n)^2$ is the right-endpoint Riemann sum for x^2 on $[0,1]$ with slice width $\frac{1}{n}\to 0$.

Chapter Seven

The Triangle That Runs Everything

One triangle of numbers, built by the simplest rule imaginable, secretly runs all of counting.

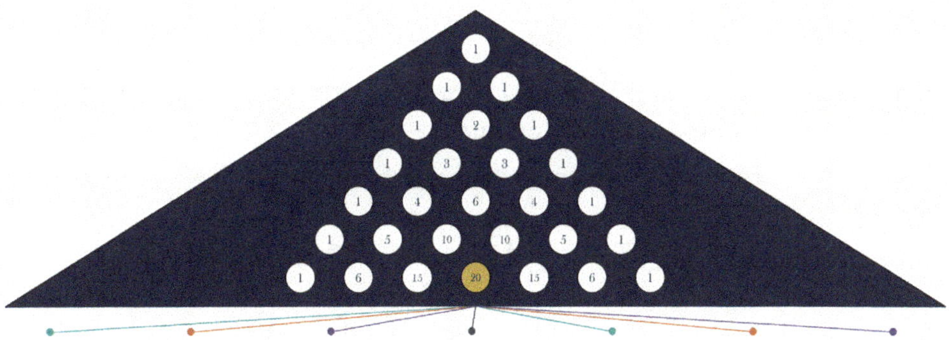

Pascal's triangle: choose, and choose again.

The next question sounds like it needs a very large piece of paper.

WONDER THIS

What are the *last two digits* of 11^{100}?

That's 11 multiplied by itself one hundred times: an enormous number. But we only want its final two digits. **Can you find them without computing the whole thing?**

CURIOUS CARL • FIRST GUESS

Carl: "If someone wants the coefficient of x^{50} in $(1+x)^{100}$, easy — I'll *expand the whole thing* and read it off!"

The catch: that's a hundred-fold multiplication for a single number, Carl — but the coefficient is just $\binom{100}{50}$, a plain *choice count*, no expansion required.

Multiplying out 11^{100} directly is hopeless by hand. But the number 11 has a gift hidden in it: $11 = 1+10$. So $11^{100} = (1+10)^{100}$, and *that* is something we can expand, provided we know how to expand a power of a sum. The tool for that is the binomial theorem, and the punchline is that we'll only need its first two terms.

When you expand $(1 + 10)^{100}$, you get a sum of pieces, each a product of some 1s and some 10s. The first few pieces are

$$(1+10)^{100} = \underbrace{1}_{\text{no tens}} + \underbrace{100 \cdot 10}_{\text{one ten}} + \underbrace{\binom{100}{2} \cdot 10^2}_{\text{two tens}} + \cdots$$

Now here is the magic for "last two digits": anything with $10^2 = 100$ or higher as a factor ends in 00: it contributes *nothing* to the last two digits. So every term from the third onward vanishes mod 100. Only the first two survive:

$$1 + 100 \cdot 10 = 1 + 1000.$$

And 1000 ends in 00, so even the second term contributes nothing to the final two digits! We're left with just 1. **The last two digits of 11^{100} are 01.** An astronomically large number, and its tail is almost all zeros: because the binomial expansion let us throw away everything that didn't matter.

The thing that made this work is the *binomial theorem*: a formula for expanding $(a + b)^n$ where the coefficients are exactly the entries of *Pascal's triangle*.

THE MOVE

The Binomial Theorem. For any n,

$$(a+b)^n = \sum_{k=0}^{n} \binom{n}{k} a^{n-k} b^k = \binom{n}{0} a^n + \binom{n}{1} a^{n-1} b + \cdots + \binom{n}{n} b^n.$$

The coefficient $\binom{n}{k}$ counts the ways to choose which k of the n factors contribute a b. So the algebra of expanding a power and the combinatorics of choosing a subset are *the same thing* — and that bridge is what makes the theorem so powerful.

That last sentence is the secret of the whole chapter. Every **binomial coefficient** is simultaneously an *algebraic* object (a coefficient in an expansion) and a *combinatorial* one (a count of subsets). Whenever you're stuck on one side, you can cross the bridge to the other.

A name for the bridge: the coefficient extractor. Write $[x^k]\,P(x)$ for "the coefficient of x^k in $P(x)$." The binomial theorem is then simply $[x^k](1+x)^n = \binom{n}{k}$, and many identities collapse into a one-line coefficient read: the whole-row sum, for instance, is $\sum_k \binom{n}{k} = (1+1)^n = 2^n$, since setting $x = 1$ just adds up every coefficient at once. Keep this little bracket in view: the showpiece below reads $\binom{2n}{n}$

straight off a *product* of two rows, and **Chapter 8** will hang whole sequences on polynomials and turn every count into a coefficient.

Pascal's triangle: the simplest rule, the richest object

Write the binomial coefficients in a triangle, row n holding $\binom{n}{0}, \binom{n}{1}, \dots, \binom{n}{n}$:

$$\begin{array}{ccccccccccc} &&&&&1&&&&& \\ &&&&1&&1&&&& \\ &&&1&&2&&1&&& \\ &&1&&3&&3&&1&& \\ &1&&4&&6&&4&&1& \\ 1&&5&&10&&10&&5&&1 \end{array}$$

$3 + 3 = 6$
each entry is the
sum of the two above

The rule that builds it is almost too simple to believe: *each entry is the sum of the two directly above it*. The two 3s in row 3 add to the 6 in row 4. That rule is **Pascal's identity**:

$$\binom{n}{k} = \binom{n-1}{k-1} + \binom{n-1}{k}.$$

And it has a one-line combinatorial proof that shows the bridge in action.

WATCH THIS #1 Why Pascal's rule is true

Claim. $\binom{n}{k} = \binom{n-1}{k-1} + \binom{n-1}{k}$.

The count. The left side counts the k-element subsets of $\{1, 2, \dots, n\}$. Now sort those subsets by a single question: *does this subset contain the element n?*

- **Yes, it contains n.** Then the other $k-1$ elements come from $\{1, \dots, n-1\}$: that's $\binom{n-1}{k-1}$ subsets.
- **No, it doesn't.** Then all k elements come from $\{1, \dots, n-1\}$: that's $\binom{n-1}{k}$ subsets.

Every k-subset falls into exactly one case, so the two add up to the whole:

$$\binom{n}{k} = \binom{n-1}{k-1} + \binom{n-1}{k}. \blacksquare$$

Trigger An identity relating binomial coefficients, or a count you can split by a yes/no question about one distinguished element.

Do first Cross the bridge: interpret each coefficient as "number of subsets," then classify the subsets by whether they contain a chosen element.

Trapdoor Trying to prove coefficient identities by grinding factorials. The combinatorial story is usually shorter and always more illuminating.

Upgrade "Does it contain element n?" generalizes to "how many of a special group does it contain?" — the engine behind Vandermonde's identity, coming up next.

That "split by a yes/no question" move is the same double-counting instinct from Chapter 1, now aimed at the triangle itself. Pascal's triangle isn't a table to memorize — it's a structure you can *rederive* any time, because every entry is just a count of subsets, and counts of subsets obey simple stories.

Reading identities off the triangle

Once you see binomial coefficients as subset-counts, a whole catalog of identities becomes obvious. Here's one with a memorable name and a beautiful diagonal picture.

WATCH THIS **#2 The hockey stick**

Claim. Summing down a diagonal, $\binom{r}{r} + \binom{r+1}{r} + \binom{r+2}{r} + \cdots + \binom{n}{r} = \binom{n+1}{r+1}$.

The count. The right side counts $(r+1)$-element subsets of $\{1, \ldots, n+1\}$. Classify each such subset by its *largest* element. If the largest element is $m+1$, the remaining r elements are chosen from $\{1, \ldots, m\}$ — that's $\binom{m}{r}$ ways. Letting m range over all its possible values sweeps out exactly the diagonal sum on the left.

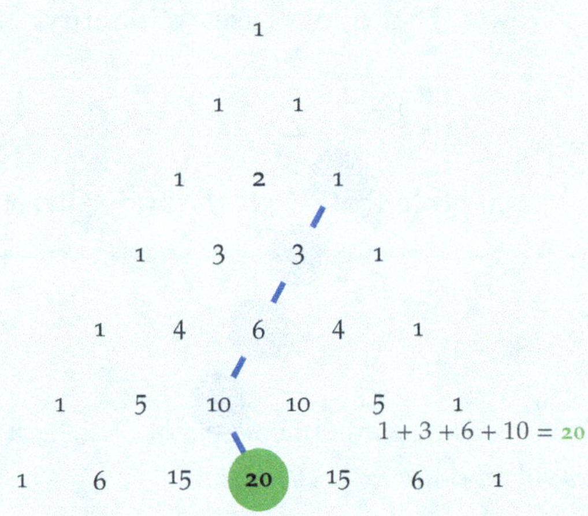

The summed diagonal forms the long handle, and the single answer cell — where the diagonal turns and bends away — forms the blade, together tracing the shape of a hockey stick. Hence the name. ■

The pattern is always the same: a binomial identity that looks like it needs algebra dissolves the instant you ask "what subsets does each side count, and how can I classify them?" Classify by the largest element, by whether a special element is included, by how many come from a certain group. Each classification is a different identity waiting to be read off.

△ TRAPDOOR

Pascal's triangle tempts you to argue by "look at the pattern" — but a pattern in the first five rows is not a proof for all rows. The combinatorial stories (classify the subsets) *are* proofs, valid for every n at once. When you spot a pattern in the triangle, don't just trust it — find the subset-counting argument that guarantees it forever.

CURIOUS CARL • TRAPDOOR!

Faced with a binomial identity, Carl rolls up his sleeves to grind: "I'll write every $\binom{n}{k}$ as a pile of factorials and force the algebra to collapse!"

A little $\binom{n}{k}$ turns around, wearing two name tags — *coefficient* on one lapel, *committee count* on the other: *"You only see my algebra tag. Read the other one and you can just* count *the committees — the identity is one honest sentence, not a page of factorials."*

The showpiece: squaring the row

Here's an identity that looks genuinely surprising the first time you see it, and whose proof is a small masterpiece of the combinatorial bridge.

WONDER THIS

Take a row of Pascal's triangle, *square every entry*, and add them up. For row 5: $1^2 + 5^2 + 10^2 + 10^2 + 5^2 + 1^2$.

Where does this land? It seems like it could be anything — but it's always a single, specific binomial coefficient. Which one?

Compute row 5: $1 + 25 + 100 + 100 + 25 + 1 = 252$. And $252 = \binom{10}{5}$, smack in the middle of row 10. That's the claim: the sum of the squares of row n equals $\binom{2n}{n}$, the central entry of row $2n$. The proof is a gem.

WATCH THIS **#3 Squaring the row, by matching coefficients**

Claim. $\sum_{k=0}^{n} \binom{n}{k}^2 = \binom{2n}{n}$.

One polynomial, read two ways. Split $2n$ equal factors down the middle:

$$(1+x)^{2n} = (1+x)^n\,(1+x)^n.$$

On the *left*, the binomial theorem gives the coefficient of x^n at a glance: it is $\binom{2n}{n}$.
On the right, multiply the two expansions $\left(\sum_i \binom{n}{i} x^i\right)\left(\sum_j \binom{n}{j} x^j\right)$. A term x^n is born whenever $i + j = n$, contributing $\binom{n}{i}\binom{n}{j} = \binom{n}{i}\binom{n}{n-i}$. Summing over i, the coefficient of x^n on the right is

$$\sum_{i=0}^{n} \binom{n}{i}\binom{n}{n-i} = \sum_{i=0}^{n} \binom{n}{i}^2,$$

where the last step uses the mirror symmetry $\binom{n}{n-i} = \binom{n}{i}$. The same coefficient computed two ways must agree, so $\sum_k \binom{n}{k}^2 = \binom{2n}{n}$. ■

Chapter 1 proved this very identity by a committee double-count: choosing n people from two rooms of n. Here the binomial theorem hands us a *second* proof, purely algebraic: the squares appear because two copies of the polynomial multiply, and the mirror symmetry $\binom{n}{n-k} = \binom{n}{k}$ folds the cross-terms into squares. *One identity, two honest proofs*: the combinatorial story of Chapter 1 and the coefficient-matching of this one. And the coefficient method generalizes for free: comparing the coefficient of x^r in $(1+x)^m(1+x)^n = (1+x)^{m+n}$ gives *Vandermonde's identity*, $\sum_k \binom{m}{k}\binom{n}{r-k} = \binom{m+n}{r}$, in full.

THE TURN *from grind to insight*

The head-on approach to $\sum_k \binom{n}{k}^2 = \binom{2n}{n}$ is to grind the algebra: write every $\binom{n}{k}$ as a ratio of factorials, square it, sum the pile, and pray the factorials collapse into the right central coefficient. It's a swamp, and even when it works it tells you *nothing* about why the answer is what it is. **The creative turn** is to see both sides as *the same coefficient* of one polynomial: since $(1+x)^{2n}$ is $(1+x)^n$ times itself, its central coefficient $\binom{2n}{n}$ must equal the sum of cross-products $\binom{n}{i}\binom{n}{n-i} = \binom{n}{i}^2$ that multiplying the two rows produces. You never manipulate a coefficient — you read one number off two factorings of a single polynomial. (Chapter 1 told the same story with committees; the binomial theorem tells it with algebra.)

TRY TO BREAK IT

Pascal's rule says every entry is the sum of the two just above it — so test it where it looks like it *shouldn't* hold: the border 1's. Take $\binom{n}{0} = 1$ at the left edge and try to write it as $\binom{n-1}{-1} + \binom{n-1}{0}$. One of those neighbors sits off the edge of the triangle entirely — so what value must $\binom{n-1}{-1}$ carry for the rule to survive at the border, and why is defining $\binom{m}{k} = 0$ for $k < 0$ or $k > m$ exactly the convention that keeps the whole triangle honest?

SEEN BEFORE?

The "count the committee two ways" move is pure **Chapter 1** (Double Counting), and the "classify by a yes/no question" proofs are its close cousins. The binomial coefficients themselves will reappear as the engine of **Chapter 8** (Generating Functions), where $(1+x)^n$ becomes a bookkeeping device, and in **Chapter 18** (Probability), where $\binom{n}{k}$ counts the ways an event can happen.

Payoff: the whole row, and the alternating row

Two famous sums fall straight out of the binomial theorem, with no work at all. You just choose what to plug in.

The binomial theorem says $(a+b)^n = \sum_k \binom{n}{k} a^{n-k} b^k$. *Set* $a = b = 1$. The left side becomes 2^n, and the right side becomes $\sum_k \binom{n}{k}$, the sum of an entire row. So **every row of Pascal's triangle sums to a power of two**: $1 + 6 + 15 + 20 + 15 + 6 + 1 = 64 = 2^6$. (We met this in Chapter 4 as "a set of n things has 2^n subsets." Here it is again, popping out of the algebra.)

Now *set* $a = 1, b = -1$. The left side becomes $(1-1)^n = 0$, and the right becomes $\sum_k \binom{n}{k}(-1)^k$ — the row with *alternating signs*. So the alternating row sum is always zero: $1 - 6 + 15 - 20 + 15 - 6 + 1 = 0$. *The same row that adds up to a power of two cancels itself to nothing, depending only on whether you wave a minus sign at it*. This is exactly the "even subsets equal odd subsets" fact from Chapter 4 — now it falls out of plugging -1 into a polynomial. The bridge between algebra and counting, paying dividends in both directions.

GOING FURTHER *optional • beyond AIME*

This detour is optional and reaches past the contest syllabus — skip it on a first pass. Pascal's triangle hides one more secret, visible if you shade every *odd* entry dark: a fractal of triangles inside triangles appears (the Sierpiński pattern). Divisibility of $\binom{m}{n}$ by a prime is not random — it is governed by the *base-p digits* of m and n, a fact known as *Lucas' theorem*. Write m and n in base p as $m = (m_d \dots m_1 m_0)_p$ and $n = (n_d \dots n_1 n_0)_p$. Then

$$\binom{m}{n} \equiv \prod_{i=0}^{d} \binom{m_i}{n_i} \pmod{p},$$

the product of the "digit-wise" binomial coefficients. If any digit n_i exceeds the matching digit m_i, that factor $\binom{m_i}{n_i}$ is 0, and the whole product collapses to 0 — so a single bad digit makes $\binom{m}{n}$ divisible by p.

A clean corollary falls right out. Take $m = p$: in base p that is $(1\,0)_p$. For $0 < k < p$, the number k is the single digit $(k)_p$, lined up as $(0\,k)_p$, so Lucas gives $\binom{p}{k} \equiv \binom{1}{0}\binom{0}{k} \equiv 0 \pmod{p}$ because $\binom{0}{k} = 0$. Thus every interior entry of row p is divisible by p — which is exactly why $(a+b)^p \equiv a^p + b^p \pmod{p}$, the "freshman's dream" behind Fermat's little theorem.

For a worked example, find $\binom{100}{50}$ mod 3. In base 3, $100 = (1\,0\,2\,0\,1)_3$ and $50 = (0\,1\,2\,1\,2)_3$, so

$$\binom{100}{50} \equiv \binom{1}{0}\binom{0}{1}\binom{2}{2}\binom{0}{1}\binom{1}{2} \pmod{3}.$$

The factor $\binom{0}{1} = 0$ (its lower index exceeds the upper), so the product collapses and $\binom{100}{50} \equiv 0 \pmod{3}$. The base-3 spelling of 50 has a digit larger than the matching digit of 100, and that single mismatch forces divisibility by 3 — the same place-value thinking as number bases, now reading the arithmetic of the triangle.

GOING FURTHER *optional • beyond AIME*

Also optional and beyond the contest syllabus. Everything so far assumed the exponent n is a positive whole number, so the expansion stops after finitely many terms. In the winter of 1665 the young Isaac Newton asked what happens when the exponent is *not* a whole number — what, for instance, is $(1+x)^{1/2} = \sqrt{1+x}$? — and his answer became one of the seeds of calculus. His move was to take the binomial-coefficient formula and refuse to let a fractional exponent stop him. For a whole number n,

$$\binom{n}{k} = \frac{n(n-1)\cdots(n-k+1)}{k!},$$

and that numerator — just k descending factors starting from n — makes perfect sense for *any* exponent. So define, for any real α, $\binom{\alpha}{k} = \frac{\alpha(\alpha-1)\cdots(\alpha-k+1)}{k!}$. For a whole number the product eventually hits a zero factor and terminates; for a fraction it never does, and Newton conjectured the expansion simply runs forever:

$$(1+x)^\alpha = \sum_{k=0}^{\infty} \binom{\alpha}{k} x^k = 1 + \alpha x + \frac{\alpha(\alpha-1)}{2}x^2 + \cdots,$$

the *generalized binomial theorem*, valid for $|x| < 1$.
Two examples show its reach. Setting $\alpha = \frac{1}{2}$ gives $\sqrt{1+x} = 1+\frac{1}{2}x-\frac{1}{8}x^2+\frac{1}{16}x^3-\cdots$; at $x = 0.21$ the first few terms already give $\sqrt{1.21} \approx 1.1000$ to five places. And setting $\alpha = -\frac{1}{2}$ with x replaced by $-x^2$ gives

$$\frac{1}{\sqrt{1-x^2}} = 1 + \tfrac{1}{2}x^2 + \tfrac{3}{8}x^4 + \cdots,$$

which Newton recognized as the slope along a circle and *integrated term by term* — each x^{2k} becoming $\frac{x^{2k+1}}{2k+1}$ — to obtain the series for the arcsine:

$$\arcsin x = x + \tfrac{1}{6}x^3 + \tfrac{3}{40}x^5 + \cdots.$$

That single manoeuvre — expand a function as an infinite power series, then integrate or differentiate it one term at a time — is the beating heart of what became calculus. The negative-exponent case $(1+x)^{-1} = 1 - x + x^2 - \cdots$ is the very geometric series that anchors Chapter 8's generating functions, but the series only converges for $|x| < 1$: plugging in $x = 3$ to compute $\sqrt{4}$ this way gives nonsense, because the terms grow instead of shrink.

What to try when you're stuck

When a problem features powers of a sum, binomial coefficients, or sums along Pascal's triangle:

1. **Expand with the binomial theorem** when you see $(a+b)^n$ — and ask which terms actually matter (as in the 11^{100} trick, where high powers vanished).
2. **Cross the bridge.** Read each $\binom{n}{k}$ as "ways to choose k from n," turning an algebra problem into a counting story.
3. **To prove an identity,** find what both sides count, then classify the objects by a clever question (largest element? contains element n? how many from group A?).
4. **Plug in clever values.** Setting a, b to $1, -1$, or other constants collapses the binomial theorem into row sums and other identities.
5. **Use Pascal's rule** $\binom{n}{k} = \binom{n-1}{k-1} + \binom{n-1}{k}$ to induct or to simplify.

THE MOVE

One-page recap. The binomial theorem $(a+b)^n = \sum_k \binom{n}{k} a^{n-k} b^k$ ties algebra to counting: $\binom{n}{k}$ is both a coefficient and a subset-count. Toolbox: **expand and discard small terms** (11^{100} mod

100), **Pascal's rule** (each entry is the sum of the two above), **classify subsets** (hockey stick, by largest element), **count two ways** (squared row = $\binom{2n}{n}$), and **plug in** ± 1 (row sums 2^n and 0). When stuck, cross the bridge between the coefficient and the count. (Two optional detours — Lucas' theorem mod p and Newton's generalized series for non-integer exponents — live in the *Going Further* boxes above.)

THE NAME FOR IT

The Binomial Theorem and Pascal's triangle

The expansion $(a+b)^n = \sum_k \binom{n}{k} a^{n-k} b^k$ is the *Binomial Theorem*, its coefficients the *binomial coefficients* arranged in *Pascal's triangle*. The committee-from-two-rooms identity is *Vandermonde's identity*, and summing down a diagonal is the *hockey-stick identity*.

WHERE IT LEADS The same coefficients reach much further. Allowing any real exponent gives Newton's *generalized (infinite) binomial theorem*, an early *Taylor / power series* and one of the historical seeds of *calculus* — the foundation of *real and complex analysis*. Reducing $\binom{m}{n}$ modulo a prime is governed by *Lucas' theorem*, opening onto the *number theory* of Pascal's triangle (Kummer's theorem and beyond). Both ideas appear in the optional *Going Further* boxes of this chapter.

CURIOUS CARL • EUREKA

Carl taps one entry on the triangle and both its name tags light up at once — "coefficient" and "committee." "You've been working two jobs this whole time, and nobody thought to tell the algebra."

SAME BEAST, NEW COSTUME

The same "it's just a choice count" beast, in costume across the book:

Subset count → Committee-with-captain → Vandermonde → Hockey stick → Roots-of-unity filter.

Each is really one binomial coefficient counting a subset, so the identity is read off two honest descriptions of the same collection rather than ground out of factorials.

The practice ladder

THE PRACTICE LADDER — *climb at your own pace*

LAUNCH *Warm the engine. Nudges below.*

1. Use the binomial theorem to find the coefficient of x^3 in $(1+x)^7$.

2. Find the coefficient of x^3 in $(2+x)^6$. (Watch the powers of 2: the term is $\binom{6}{k}2^{6-k}x^k$.)

CORE BUILDER *Master the central moves.*

1. Prove $\sum_{k=0}^{n} \binom{n}{k} = 2^n$ both by plugging into the binomial theorem and by counting subsets.

2. Prove $\sum_{k=0}^{n} k\binom{n}{k} = n\,2^{n-1}$ (you met the committee-with-a-captain version in Chapter 1 — now derive it by differentiating $(1+x)^n$, or by a coefficient argument).

3. Use the hockey-stick identity to evaluate $\binom{2}{2} + \binom{3}{2} + \binom{4}{2} + \cdots + \binom{10}{2}$.

STRETCH *Transfer to unfamiliar ground.*

1. Prove Vandermonde's identity $\sum_k \binom{m}{k}\binom{n}{r-k} = \binom{m+n}{r}$ by the two-rooms committee argument.

2. Find the last two digits of 3^{100} using $3^{100} = 9^{50} = (10-1)^{50}$.

3. Find the constant term (the term with no x) in the expansion of $\left(x + \frac{2}{x}\right)^6$. (Write the general term and ask which power of x vanishes.)

THE FLASH *No clock — each cracks open in one flash of insight, once you find the angle. Nudges at the back.*

1. What is the sum of all the entries in the first $n+1$ rows of Pascal's triangle (rows 0 through n)?

2. Find the coefficient of x^4 in $(1+2x)^5$.

PROOF LAB *Write it clean, in full sentences.*

1. Prove the binomial theorem $(a+b)^n = \sum_{k=0}^{n} \binom{n}{k} a^{n-k} b^k$ by a counting argument: expanding the product means choosing a or b from each of the n factors.

2. Prove $\sum_{k=0}^{n} \binom{n}{k}^2 = \binom{2n}{n}$ and state the general Vandermonde identity it specializes.

REFLECTION *No nudge. Write a few sentences.*

1. For one identity above, did you prove it by algebra (plugging in / expanding) or by a counting story? Which felt more convincing, and why?

NUDGES

Launch. (1) $\binom{7}{3} = 35$. (2) Term is $\binom{6}{k}2^{6-k}x^k$; for x^3 take $k = 3$: $\binom{6}{3}2^3 = 20 \cdot 8 = 160$. **Core.** (3) Plug $a = b = 1$; a set of n elements has 2^n subsets. (4) Differentiate $(1 + x)^n$, then set $x = 1$. (5) Hockey stick gives $\binom{11}{3} = 165$. **Stretch.** (6) Choose r from m seniors and n juniors; split by how many come from the seniors. (7) $(10-1)^{50}$ mod 100: only $k = 0, 1$ survive, giving $1 - 500 \equiv 1$; last two digits 01. (8) General term $\binom{6}{k}2^k x^{6-2k}$; constant at $k = 3$: $\binom{6}{3}2^3 = 160$. **The Flash.** (9) $2^0 + 2^1 + \cdots + 2^n = 2^{n+1} - 1$. (10) $\binom{5}{4}2^4 = 5 \cdot 16 = 80$. **Proof.** (11) Expand $(a + b)^n$ as a product of n identical factors $(a + b)$; multiplying out means picking a or b from each factor and summing all 2^n products. A product with exactly k chosen b's equals $a^{n-k}b^k$, and there are $\binom{n}{k}$ ways to choose which factors give b; collecting like terms gives $\sum_{k=0}^{n}\binom{n}{k}a^{n-k}b^k$. (12) Count the $\binom{2n}{n}$ ways to pick n people from two rooms of n each. Taking k from the first room and $n - k$ from the second gives $\binom{n}{k}\binom{n}{n-k} = \binom{n}{k}^2$ (using $\binom{n}{n-k} = \binom{n}{k}$); summing over k counts every committee, so $\sum_{k=0}^{n}\binom{n}{k}^2 = \binom{2n}{n}$. This is Vandermonde's identity $\sum_k \binom{m}{k}\binom{r}{n-k} = \binom{m+r}{n}$ specialized to $m = r = n$.

8

A Clothesline for Numbers

Hang a whole infinite sequence on the pegs of a polynomial's powers — then let algebra do the counting for you.

Generating functions: turn counting into algebra.

This ought to be impossible.

WONDER THIS

The Fibonacci numbers $0, 1, 1, 2, 3, 5, 8, 13, \ldots$ obey the tangled rule $F_n = F_{n-1} + F_{n-2}$ — each one leans on the two before it, so to reach the hundredth you seem to need all ninety-nine earlier ones.

Could a single function hold the whole infinite sequence at once — and hand you a formula for F_n with no recursion at all?

It sounds impossible, yet the answer is one short line:

$$F(x) = \frac{x}{1 - x - x^2}$$

holds *every* Fibonacci number in its power-series coefficients, and from it Binet's exact formula for F_n falls out by pure algebra. Nothing shows off this chapter's idea better: a messy recurrence becomes a trivial fraction. By the end you will derive that fraction in three lines — and to earn it, we begin somewhere you can count on your fingers. You have an unlimited supply of pennies, nickels, and dimes; **in how many ways can you make 10 cents?** Order doesn't matter, only how many of each coin. Listing gives just *four* (ten pennies; five pennies and a nickel; two nickels; a dime) — but for a dollar, or stranger coins, listing collapses. We want a *machine* that counts for us, and it turns out to be a single polynomial.

CURIOUS CARL • FIRST GUESS

Carl: "Ten cents? Then surely *ten* ways — one per cent!"

The catch: coins don't care about order, Carl, and cents aren't choices — list them and you'll find only *four*. What we want is a machine that never miscounts, and it turns out to be a single polynomial.

Here's the idea. For the pennies, write down the polynomial $1 + x + x^2 + x^3 + \cdots$, where the exponent records how many cents we commit in pennies (we can use 0 pennies, or 1, or 2, ...). For the nickels, only multiples of 5 are possible, so write $1 + x^5 + x^{10} + \cdots$. For the dimes, $1 + x^{10} + x^{20} + \cdots$. Now *multiply these three together*. When you expand the product, a term like x^{10} arises every time you pick a penny-power, a nickel-power, and a dime-power whose exponents *add up to* 10. That is exactly a way of making 10 cents! So:

$$(\text{ways to make } n \text{ cents}) = \text{the coefficient of } x^n \text{ in } (1 + x + x^2 + \cdots)(1 + x^5 + \cdots)(1 + x^{10} + \cdots).$$

The whole counting problem became a single coefficient in a single product. The polynomial *generates* the sequence of answers, one per coefficient, so it's called a **generating function**. We turned coins into exponents, and "adding up to n" into "multiplying and collecting the x^n term."

THE MOVE

A **generating function** encodes a sequence $a_0, a_1, a_2, \ldots$ as the coefficients of a power series:

$$A(x) = a_0 + a_1 x + a_2 x^2 + a_3 x^3 + \cdots$$

The variable x is just a bookkeeping peg — you rarely plug a number into it. The magic is that *operations on sequences become operations on functions*: shifting becomes multiplying by x, combining choices becomes multiplying series, and a recurrence becomes a single solvable equation. Find the function, and the sequence is just its coefficients.

This is one of the most powerful ideas in all of combinatorics, and it can feel like sorcery the first time: you convert a counting problem into algebra, turn the algebra crank, and read the answer off the result. The sequence you wanted was hanging on the clothesline the whole time.

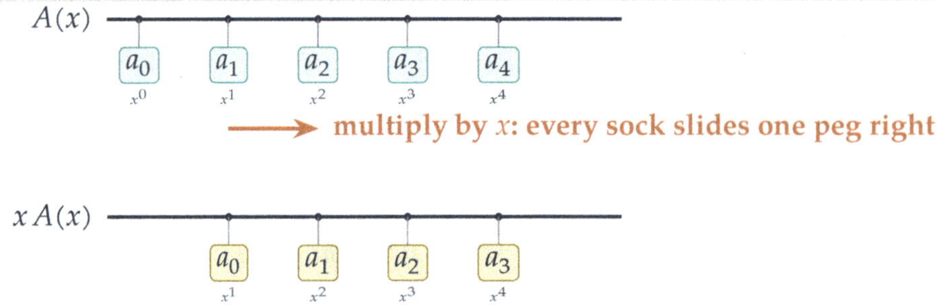

The clothesline. Pin the whole sequence to one line, each term a_k on the peg x^k. Then the operations you care about become tugs on the line: *multiplying by x slides every term one peg to the right* (a shift), and *multiplying two clotheslines* tangles their pegs so exponents add, exactly "combine choices." A recurrence, which ties each term to its neighbors, becomes a single equation the whole line must satisfy.

The two series you must know

Generating-function fluency rests on recognizing a few standard series. Two of them carry most of the load, and both come from things you already know.

The first is the *geometric series*, the single most useful identity in the subject:

$$1 + x + x^2 + x^3 + \cdots = \frac{1}{1-x}.$$

That compact right-hand side is what makes the magic practical: the infinite penny-polynomial is just $\frac{1}{1-x}$. The second is the binomial theorem from Chapter 7, which tells us how powers of $\frac{1}{1-x}$ behave.

WATCH THIS **#1 Choosing with repetition**

Claim. The number of ways to choose n items from k types, with repetition allowed, is $\binom{n+k-1}{k-1}$ — and the generating function proves it in one line.

The setup. Each of the k types can be chosen $0, 1, 2, \ldots$ times, contributing the series $1 + x + x^2 + \cdots = \frac{1}{1-x}$. With k independent types, multiply k copies:

$$\underbrace{\frac{1}{1-x} \cdot \frac{1}{1-x} \cdots \frac{1}{1-x}}_{k \text{ factors}} = \frac{1}{(1-x)^k}.$$

Read off the coefficient. A standard expansion (the binomial theorem, extended to negative powers) gives

$$\frac{1}{(1-x)^k} = \sum_{n=0}^{\infty} \binom{n+k-1}{k-1} x^n.$$

So the coefficient of x^n — the number of ways — is $\binom{n+k-1}{k-1}$. For $n = 7$ items from $k = 3$ types: $\binom{9}{2} = 36$. ■

Trigger A counting problem about combining independent choices, especially "how many ways to total n" from repeatable parts.

Do first Build a series for each independent choice (often $\frac{1}{1-x}$ or a finite piece of it), multiply, and extract the coefficient of x^n.

Trapdoor Forgetting the closed forms. $1 + x + x^2 + \cdots = \frac{1}{1-x}$ and the expansion of $\frac{1}{(1-x)^k}$ are the workhorses — not knowing them leaves you stuck with an infinite product.

Upgrade This is the stars-and-bars count from Chapter 4, now derived by algebra instead of a picture — the same answer from a completely different machine.

Notice this is the stars-and-bars result from Chapter 4 all over again. Where before we drew dots and dividers, here we just multiplied k copies of $\frac{1}{1-x}$ and read off a coefficient. That's the recurring theme: generating functions reprove combinatorial facts *mechanically*, and once the machine is built, you barely have to think. The cleverness is loaded entirely into setting up the right product.

△ TRAPDOOR

Generating functions are *formal* — x is a placeholder, and questions of "does the series converge" usually don't matter for counting. Don't get distracted trying to plug in values of x or worry about convergence; treat the series as an infinite bookkeeping device. The one real danger is misbuilding a factor: make sure each series correctly lists the *allowed* amounts for that choice (e.g. nickels contribute $1 + x^5 + x^{10} + \cdots$, not $1 + x + x^2 + \cdots$).

CURIOUS CARL • TRAPDOOR!

Carl pins up the nickel factor as $1 + x + x^2 + x^3 + \cdots$, humming happily. "Every exponent should get a peg!"

The clothesline sags and complains: *"Not every peg — only the ones a nickel can reach! Hang me at* $1, x^5, x^{10}, \ldots$, *or you'll be counting change that doesn't exist."* A factor may only list the amounts its coin is actually allowed to contribute.

The showpiece: Catalan in one equation

We've now met the Catalan numbers twice: by bijection (Chapter 4) and by recurrence (Chapter 5). The generating function delivers them a third way, and this one is the most powerful, because it produces a *formula* for all of them at once from a single self-referential equation.

WONDER THIS

Recall the Catalan recurrence from Chapter 5: $C_{n+1} = \sum_{i=0}^{n} C_i C_{n-i}$, with $C_0 = 1$.

Can we capture this entire infinite family of equations in one stroke — and solve for a closed form?

The key realization: that convolution sum $\sum_i C_i C_{n-i}$ is *exactly* the coefficient pattern you get when you *square* a generating function. This is the deep fact that makes the whole method sing.

WATCH THIS #2 A function that contains itself

Setup. Let $C(x) = C_0 + C_1x + C_2x^2 + \cdots$ be the generating function for the Catalan numbers. When you multiply $C(x)$ by itself, the coefficient of x^n in $C(x)^2$ is, by the way polynomials multiply,

$$\sum_{i=0}^{n} C_i C_{n-i}$$

— which the recurrence tells us equals C_{n+1}.

Translate the recurrence into an equation. So $C(x)^2$ is the series whose x^n coefficient is C_{n+1} — that is, the Catalan sequence shifted down by one. Shifting down by one and accounting for the starting term $C_0 = 1$:

$$C(x) = 1 + x\,C(x)^2.$$

The generating function *satisfies an equation involving itself* — a perfect mirror of the recurrence's self-reference.

recurrence $C_{n+1} = \sum C_i C_{n-i}$ —becomes→ $C = 1 + xC^2$

Solve it like algebra. Treating $C = C(x)$ as an unknown, $xC^2 - C + 1 = 0$ is just a quadratic. The quadratic formula gives

$$C(x) = \frac{1 - \sqrt{1 - 4x}}{2x}$$

(we take the minus sign so that $C(0) = C_0 = 1$). Expanding that square root by the binomial series and simplifying yields the coefficients

$$C_n = \frac{1}{n+1}\binom{2n}{n},$$

the closed form for the Catalan numbers — pulled from thin air by an equation a function wrote about itself. ■

Sit with how strange and wonderful that was. A recurrence is a rule connecting *infinitely many* numbers; we packed all of them onto one clothesline $C(x)$, and the recurrence collapsed into a *single quadratic equation*. Solving the quadratic and expanding the root handed us a formula valid for every Catalan number simultaneously. Three chapters, three lenses (bijection, recurrence, generating function) and the same sequence each time, each method illuminating something the others couldn't. That is the quiet thesis of this whole book made visible: the techniques are not separate tricks; they are different windows onto the same mathematical objects.

THE TURN *from grind to insight*

The patient route to the Catalan numbers is to obey the recurrence literally: compute C_2 from C_0, C_1, then C_3 from those, then C_4 — each step a fresh convolution sum, the arithmetic ballooning, and never a closed form in sight, only a longer and longer list of numbers. **The creative turn**

is to refuse the term-by-term march and pack the *entire* infinite sequence onto a single function $C(x)$, where the convolution becomes nothing but $C(x)^2$ and the whole recurrence shrinks to one self-referential equation $C = 1 + xC^2$. It wins because that one equation, solved like ordinary algebra, hands you *every* Catalan number at once — a formula, not a forecast you have to keep recomputing.

SEEN BEFORE?

This closes the Catalan thread from **Chapter 4** (the reflection bijection) and **Chapter 5** (the split-in-the-middle recurrence). The "square the generating function = convolve the sequence" principle is the same convolution from Chapter 5, now wearing algebra's clothes. And the binomial-series expansion of $\sqrt{1-4x}$ leans directly on **Chapter 7**.

TRY TO BREAK IT

Strip the coins down to pennies and nickels only, and write the generating function: $\frac{1}{(1-x)(1-x^5)}$. Chase the coefficient of x^{25} — how many ways to make 25 cents now? Then push harder: what if *order* matters, so that a nickel-then-penny counts differently from a penny-then-nickel (a *composition*)? Watch which factor of the machine has to change, and you'll feel exactly where "order-blind" lives inside the polynomial.

Payoff: Fibonacci, as promised

We opened this chapter with a claim: one little fraction holds every Fibonacci number, with Binet's formula waiting inside it. Now we have the tools to earn it.

Let $F(x) = F_0 + F_1x + F_2x^2 + \cdots$. The recurrence $F_n = F_{n-1} + F_{n-2}$ says that $F(x)$, minus a shifted copy of itself, minus a twice-shifted copy, collapses to almost nothing. Multiplying a generating function by x *delays* the sequence by one place (the coefficient of x^n in $xF(x)$ is F_{n-1}), and multiplying by x^2 delays it by two. Carefully assembling $F(x) - xF(x) - x^2F(x)$ and using the recurrence to cancel everything past the first term leaves

$$F(x)(1 - x - x^2) = x \qquad \Longrightarrow \qquad F(x) = \frac{x}{1 - x - x^2}.$$

Every Fibonacci number lives inside that one little fraction. If you wanted, you could factor the denominator, split into partial fractions, and out would fall *Binet's formula* — the exact expression for F_n in terms of the golden ratio φ. The recurrence's growth fingerprint we noticed back in Chapter 5 (terms scaling by φ) is literally the root of the denominator $1 - x - x^2$. *The whole infinite sequence, folded into a fraction the size of a postage stamp.*

GOING FURTHER *optional • beyond AIME*

Optional, beyond contest scope — skip this on a first pass and lose nothing the rest of the chapter depends on. Everything so far hung sequences on the pegs $1, x, x^2, x^3, \ldots$, which is perfect for *unlabeled* counting — coins, partitions, compositions — where all that matters is how many of each kind. But a huge family of problems involves *labeled* objects: people in distinct seats, permutations of distinct symbols, functions between named sets. For those there is a second clothesline, the *exponential generating function* (EGF), which hangs the sequence $a_0, a_1, a_2, \ldots$ not on x^n but on $x^n/n!$:

$$A(x) = \sum_{n\geq 0} a_n \frac{x^n}{n!} = a_0 + a_1 x + a_2 \frac{x^2}{2} + a_3 \frac{x^3}{6} + \cdots$$

The factorial weighting looks like an arbitrary tax, but it is exactly the right weight for labeled structures, and the reason is in how two EGFs multiply. If $A(x) = \sum a_n \frac{x^n}{n!}$ and $B(x) = \sum b_n \frac{x^n}{n!}$, then their product is the EGF whose n-th coefficient is

$$[x^n/n!]\ A(x)B(x) = \sum_{k=0}^{n} \binom{n}{k} a_k\, b_{n-k}.$$

The binomial coefficient is the whole point. To build a combined labeled object on n items, *choose* which k of the n labels go to the A-part ($\binom{n}{k}$ ways), build an A-structure on them (a_k ways), and a B-structure on the remaining $n-k$ (b_{n-k} ways). For ordinary GFs, multiplying merges *unlabeled* amounts ($a_k b_{n-k}$, no binomial); for EGFs it merges *labeled* ones, with the $\binom{n}{k}$ that distributes the labels coming along for free. The cleanest test case is the all-ones sequence $a_n = 1$ — "one way to do nothing special to n labeled items" — whose EGF is $\sum_{n\geq 0} \frac{x^n}{n!} = e^x$. The exponential function *is* the EGF of the all-ones sequence, which is where the method gets its name.
One merge then recovers the derangement formula — the sequence D_n from Chapter 5's recurrence practice. A *derangement* is a permutation of $\{1, \ldots, n\}$ with no fixed point; let D_n count them and $D(x) = \sum_{n\geq 0} D_n \frac{x^n}{n!}$ be its EGF. Every permutation splits cleanly into two labeled parts: its set of fixed points (each left alone) and a derangement of the rest. The fixed-point set has exactly one structure on any subset, so its EGF is the all-ones e^x; merging it with a derangement, by the product rule, builds an arbitrary permutation, giving $e^x \cdot D(x) = (\text{EGF of all permutations})$. Since there are $n!$ permutations of n items, that EGF is $\sum_{n\geq 0} n! \frac{x^n}{n!} = \sum_{n\geq 0} x^n = \frac{1}{1-x}$. Therefore $e^x D(x) = \frac{1}{1-x}$, so $D(x) = \frac{e^{-x}}{1-x}$. Multiplying the two known series $e^{-x} = \sum_k \frac{(-1)^k}{k!} x^k$ and $\frac{1}{1-x} = \sum_m x^m$, the coefficient of x^n in the product is $\sum_{k=0}^{n} \frac{(-1)^k}{k!}$; and because $D(x)$ is an EGF, $D_n = n!\,[x^n] D(x)$, so

$$D_n = n! \sum_{k=0}^{n} \frac{(-1)^k}{k!}.$$

For $n = 4$ this gives $24\left(1 - 1 + \frac{1}{2} - \frac{1}{6} + \frac{1}{24}\right) = 9$. The inclusion–exclusion count of derangements fell out in a few lines of algebra, because the EGF product rule *is* the labeled-merge logic that inclusion–exclusion tracked by hand — and the famous limit $D_n/n! \to e^{-1}$ is now obvious, since the bracket is a partial sum of the series for e^{-1}.

What to try when you're stuck

When a problem asks "how many ways to total n," or hands you a recurrence, or involves combining independent choices, reach for a generating function:

1. **Build a series for each independent choice.** What amounts can this part contribute? Write $\sum x^{(\text{amount})}$ — often $\frac{1}{1-x}$ or a finite piece.
2. **Multiply the series.** The coefficient of x^n in the product counts the ways to total n.
3. **For a recurrence,** define $A(x) = \sum a_n x^n$, translate the recurrence into an equation in $A(x)$ (shifting $\leftrightarrow$ multiplying by x; convolution $\leftrightarrow$ squaring), and solve.
4. **Know your closed forms:** $\frac{1}{1-x} = \sum x^n$ and $\frac{1}{(1-x)^k} = \sum \binom{n+k-1}{k-1} x^n$.
5. **Extract the coefficient** by expanding, by partial fractions, or by recognizing a known series.

THE MOVE

One-page recap. A generating function hangs a sequence on the coefficients of a power series; operations on sequences become algebra on functions. Toolbox: **a series per independent choice, then multiply** (making change), $\frac{1}{1-x}$ **and** $\frac{1}{(1-x)^k}$ (geometric and choose-with-repetition), **convolution = squaring** (Catalan's $C = 1 + xC^2$), and **recurrence → one equation** (Fibonacci's $\frac{x}{1-x-x^2}$). When stuck counting totals or solving a recurrence, put the whole sequence on a clothesline and do algebra.

THE NAME FOR IT

Generating functions

Encoding a sequence as the coefficients of a power series produces its *generating function*; manipulating the series algebraically is the method of *generating functions*. The workhorse identity $1 + x + x^2 + \cdots = \frac{1}{1-x}$ is the *geometric series*.

WHERE IT LEADS Generating functions are the combinatorial face of the *Z-transform* and *Fourier/Laplace transforms*: encoding a sequence as a function and doing analysis on it. They are the natural language of *probability* (moment and characteristic functions) and *analytic combinatorics*, where a function's singularities reveal a sequence's growth.

CURIOUS CARL • EUREKA

Carl pegs the entire sequence to a clothesline like mismatched socks. "I used to chase these one at a time. Now they just hang there and let the algebra read them off. Lazy. *Perfect.*"

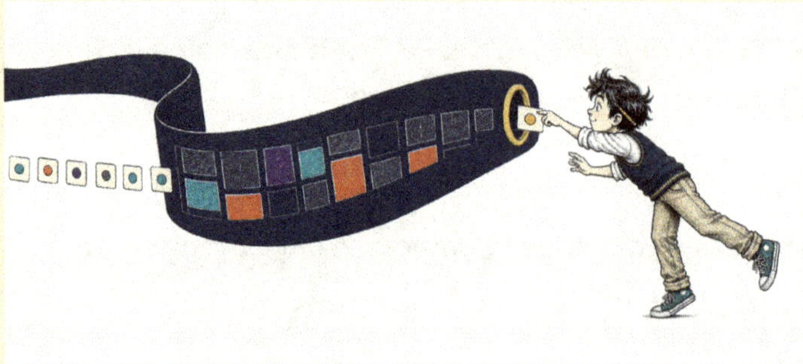

SAME BEAST, NEW COSTUME

The same "hang the sequence and multiply" beast, in disguise across the book:

Coin counting → Choose with repetition → Catalan $C = 1 + xC^2$ **→ Fibonacci GF → Partition products.**

Each is one function whose coefficients *are* the answers. Build the right product or equation, and the whole sequence falls out at once.

The practice ladder

THE PRACTICE LADDER *climb at your own pace*

LAUNCH *Warm the engine. Nudges below.*

1. Write the generating function for the sequence $1, 1, 1, 1, \ldots$ and for $1, 2, 3, 4, \ldots$ in closed form.

2. What is the coefficient of x^4 in $(1 + x + x^2 + \cdots)^2$? Interpret it as a counting answer.

CORE BUILDER *Master the central moves.*

1. Using generating functions, find the number of ways to make 25 cents with pennies, nickels, dimes, and quarters. (Set up the product; you may expand by computer or by hand.)

2. Find the generating function for the sequence $a_n = a_{n-1} + 2$, $a_0 = 1$, and extract a closed form for a_n.

3. Show that the number of ways to write n as an ordered sum of 1s and 2s has generating function $\frac{1}{1-x-x^2}$, and connect it to Fibonacci.

STRETCH *Transfer to unfamiliar ground.*

1. Find the generating function for the number of partitions of n into parts of size at most 3, and use it to compute the count for $n = 6$.

2. Solve $a_n = 3a_{n-1} - 2a_{n-2}$, $a_0 = 0$, $a_1 = 1$, by finding $A(x)$ as a rational function and decomposing into partial fractions.

THE FLASH *No clock — each cracks open in one flash of insight, once you find the angle. Nudges at the back.*

1. What is the coefficient of x^7 in $\frac{1}{(1-x)^3}$?

2. Write the generating function for $0, 1, 2, 3, \ldots$ (the non-negative integers).

PROOF LAB *Write it clean, in full sentences.*

1. Derive the Catalan generating function from $C = 1 + xC^2$, choosing the correct branch of the quadratic, and explain why $C(0) = 1$ forces the sign.

2. Prove that the generating function for a sequence satisfying $a_n = a_{n-1} + a_{n-2}$ (with $a_0 = 0, a_1 = 1$) is $\frac{x}{1-x-x^2}$, justifying the shift-equals-multiply-by-x step.

REFLECTION *No nudge. Write a few sentences.*

1. The Catalan numbers have now appeared via bijection, recurrence, and generating function. Which of the three felt most illuminating to you, and what did it reveal that the others didn't?

NUDGES

Launch. (1) $\frac{1}{1-x}$ and $\frac{1}{(1-x)^2}$. (2) Coefficient 5: the ways to write $4 = i + j$ with $i, j \geq 0$. **Core.** (3) Coefficient of x^{25} in $\frac{1}{(1-x)(1-x^5)(1-x^{10})(1-x^{25})}$, which is 13. (4) $A(x) = \frac{1}{1-x} + \frac{2x}{(1-x)^2}$; closed form $a_n = 2n + 1$. (5) Each composition into 1s and 2s gives a factor; the series is $\frac{1}{1-(x+x^2)}$. **Stretch.** (6) $\frac{1}{(1-x)(1-x^2)(1-x^3)}$; coefficient of x^6 is 7. (7) $A(x) = \frac{x}{(1-x)(1-2x)}$; partial fractions give $a_n = 2^n - 1$. **The Flash.** (8) $\binom{9}{2} = 36$. (9) $\frac{x}{(1-x)^2}$. **Proof.** (10) Rearranging $C = 1 + xC^2$ gives $xC^2 - C + 1 = 0$, so $C = \frac{1 \pm \sqrt{1-4x}}{2x}$. As $x \to 0$ the $+$ branch has numerator $\to 2$ and blows up, contradicting $C(0) = 1$; expanding $\sqrt{1-4x} = 1 - 2x - 2x^2 - \cdots$ shows the $-$ branch $\to 1$. Hence $C = \frac{1-\sqrt{1-4x}}{2x}$. (11) Let $F(x) = \sum_{n \geq 0} a_n x^n$. Multiplying by x shifts indices, so $xF(x) = \sum a_{n-1} x^n$ and $x^2 F(x) = \sum a_{n-2} x^n$; then $F(x) - xF(x) - x^2 F(x) = \sum_n (a_n - a_{n-1} - a_{n-2}) x^n$. The recurrence kills every coefficient for $n \geq 2$, leaving only $a_0 + (a_1 - a_0)x = x$. Thus $F(x)(1 - x - x^2) = x$, i.e. $F(x) = \frac{x}{1-x-x^2}$.

MATH LAB #2

Catalan Carnival

One sequence, a dozen costumes. Walk the midway, match the disguises, and watch the same numbers — 1, 1, 2, 5, 14, ... — turn up at every booth.

In Chapters 4, 5, and 8 the *Catalan numbers* kept crashing the party: as ticket-counter paths, as parenthesizations, as triangulations, and as the solution of $C(x) = 1 + xC(x)^2$. This lab is the reunion. At four booths you'll meet the *same* count wearing four masks — tune each booth to $n = 3$ and you should see **five** of everything. That is no coincidence; it's the whole point.

CURIOUS CARL • FIRST GUESS

Carl: "Four different objects? Then four different answers — I'll count each booth from scratch!"

The catch: count *one* booth honestly, Carl, then look for the *pairing* that carries your answer to the next. The carnival's trick is that every booth is secretly the same booth.

STATION 1 • the Parentheses booth

List every way to fully parenthesize the product $a\,b\,c\,d$ (four factors, multiplied two at a time). **How many are there?** Then draw each as a *binary tree* whose leaves are a, b, c, d in order and whose root is the *last* multiplication.

STATION 2 • the Mountain-Path booth

A path takes 3 up-steps (↗) and 3 down-steps (↘) and *never dips below where it started.* **How many such mountain paths are there?** Match each to a string of 3 pairs of balanced parentheses (up = "(", down = ")").

STATION 3 • the Triangulation booth

Take a convex *pentagon* and cut it into triangles using non-crossing diagonals. **How many triangulations are there?** (Fix one edge as the base and split on the triangle sitting on it.)

STATION 4 • the Machine booth

Every booth obeys the convolution recurrence $C_{n+1} = \sum_{i=0}^{n} C_i C_{n-i}$ with $C_0 = 1$. Package the whole sequence into $C(x) = \sum_{n\geq 0} C_n x^n$; the recurrence becomes the single equation

$$C(x) = 1 + x\,C(x)^2.$$

Solve this quadratic for $C(x)$, choose the root with $C(0) = 1$, and expand to read off $C_0, \dots, C_4$.

STATION 5 • Boss — prove two booths are the same

Pick the Mountain-Path booth and the Parentheses booth. **Build an explicit bijection** between 3-pair balanced-parenthesis strings and mountain paths of 3 ups and 3 downs, and check it is reversible. (You've then *proved* the two counts are equal without recomputing either.)

NUDGES

Peeks (try first!). **(1)** $\boxed{5}$: $((ab)c)d$, $(a(bc))d$, $(ab)(cd)$, $a((bc)d)$, $a(b(cd))$ — and five matching binary trees ($= C_3$). **(2)** $\boxed{5}$ paths ↔ the five strings ((())), (()()), (())(), ()(()), ()()(). **(3)** $\boxed{5}$ pentagon triangulations (one per choice of the two diagonals) $= C_3$. **(4)** $xC^2 - C + 1 = 0 \Rightarrow C(x) = \dfrac{1-\sqrt{1-4x}}{2x}$, whose expansion is $1 + x + 2x^2 + 5x^3 + 14x^4 + \cdots$, so $C_0..C_4 = 1, 1, 2, 5, 14$. **(5)** Read the string left to right: send "(" to an up-step and ")" to a down-step. "Balanced" (never more ")" than "(" so far) is *exactly* "never dips below start," and the map is undone by reading the path back into brackets — a bijection.

CURIOUS CARL • EUREKA

Carl: "Oh — it was *five* at every booth because the booths are the same ride in different paint! Count once, then just translate."

CHAPTER NINE

9

Only the Remainder Matters

Stop caring about numbers. Start caring only about their remainders — and watch hard problems become easy.

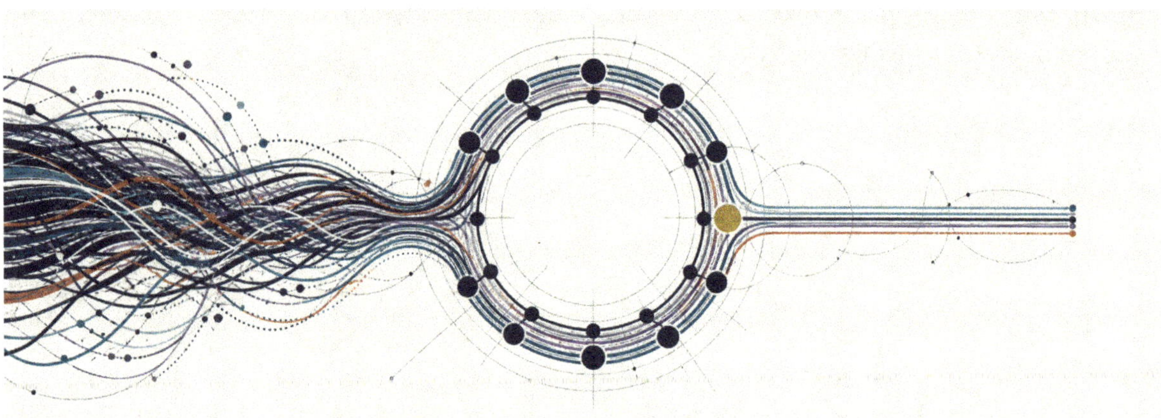

Modular arithmetic: keep only the remainder.

Here's a trick you can perform on anyone.

WONDER THIS

Ask a friend to pick any three-digit number — say 274 — and write it down *twice in a row* to make a six-digit number: 274274.

Now tell them, without seeing it, that their number is divisible by 7, by 11, and by 13. It always is. **Why?**

It feels like sorcery, but it's pure structure. Writing 274 twice in a row isn't a random act — it's an arithmetic operation. Sticking a copy of 274 in front of another copy multiplies the first by 1000 and adds the second:

$$274274 = 274 \times 1000 + 274 = 274 \times 1001.$$

And here's the secret hiding in plain sight: $1001 = 7 \times 11 \times 13$. So *every* number of the form $\overline{abcabc}$ is $\overline{abc} \times 1001$, automatically carrying 7, 11, and 13 as factors. **The trick works because 1001 factors into exactly those three primes** — and repeating a three-digit block is secretly multiplying by 1001. So you refuse to compute it: keep only what survives division, and the astronomical becomes a clock you can read. The infinite is not tamed by bigger effort, but by the nerve to throw almost all of it away and trust what little remains.

That's the spirit of this chapter. Instead of wrestling with gigantic numbers, we ask what they're *divisible* by, or what *remainder* they leave — and those questions have clean, structured answers. The central tool is a shift in attitude called **modular arithmetic**: working with remainders as if they were numbers in their own right.

THE MOVE

We write $a \equiv b \pmod{m}$ ("a is congruent to b mod m") to mean a and b leave the *same remainder* when divided by m — equivalently, m divides $a - b$. The miracle is that congruences *add, subtract, and multiply* just like equations: if $a \equiv b$ and $c \equiv d \pmod{m}$, then $a + c \equiv b + d$ and $ac \equiv bd \pmod{m}$. So you can replace any number by its remainder *at any point* in a calculation, keeping the arithmetic tiny.

CURIOUS CARL • FIRST GUESS

Carl: "The last digit of 7^{2024}? I compute 7^{2024} and read the end off." (*He reaches for a very long sheet of paper.*)

The catch: you only want the *last* digit, Carl — so work mod 10, where the powers of 7 cycle $7, 9, 3, 1$ and you need only track the remainder, never the monster.

That last property is what makes modular arithmetic so powerful: you never have to carry around a huge number when all you care about is its remainder. You can reduce early, reduce often, and keep every quantity small. Whole calculations that would overflow a page collapse into single-digit arithmetic.

Why the divisibility rules work

You probably learned tricks in grade school: "a number is divisible by 3 if its digits sum to a multiple of 3," "divisible by 9 likewise," "divisible by 11 if the alternating digit sum works out." These aren't arbitrary — each one is a one-line consequence of reducing powers of 10 modulo the divisor.

WATCH THIS #1 Casting out nines

Claim. A number is divisible by 9 exactly when the sum of its digits is. (And in general, a number is congruent to its digit sum, mod 9.)

The key fact. Modulo 9, every power of ten is just 1:

$$10 \equiv 1, \quad 100 \equiv 1, \quad 1000 \equiv 1, \quad \dots \quad \pmod 9,$$

because $10 = 9 + 1 \equiv 1$, and multiplying 1 by itself stays 1.

The consequence. Write a number by its digits, say $\overline{d_k \cdots d_1 d_0} = d_k 10^k + \cdots + d_1 10 + d_0$. Reduce each power of ten to 1:

$$\overline{d_k \cdots d_0} \equiv d_k \cdot 1 + \cdots + d_1 \cdot 1 + d_0 = (d_k + \cdots + d_1 + d_0) \pmod 9.$$

So a number and its digit sum are congruent mod 9 — in particular, one is divisible by 9 exactly when the other is. For 1234: digit sum $10 \equiv 1 \pmod 9$, and indeed $1234 = 137 \times 9 + 1$. ■

Trigger A divisibility rule or a question about remainders of a number written in decimal.

Do first Reduce each power of 10 modulo the divisor, then replace the number by the simpler combination of its digits.

Trapdoor Forgetting that the rule depends on what 10 reduces to: mod 9 it's 1 (plain digit sum), but mod 11 it's -1 (*alternating* digit sum).

Upgrade The same "reduce the base" idea gives rules for 11 ($10 \equiv -1$, so alternating sum), 7, 13, and the 1001 trick from the opener.

The rule for 11 is the same move with one twist, and it's clean enough to just pocket.

THE MOVE

Divisibility by 11. Since $10 \equiv -1 \pmod{11}$, powers of 10 alternate: $1, -1, 1, -1, \dots$. So a number is $\equiv$ its *alternating digit sum* $\pmod{11}$ — add the digits in odd places, subtract those in even places, and check that against 11. (E.g. 9174: $4 - 7 + 1 - 9 = -11 \equiv 0$, so $11 \mid 9174$.)

That one question hides behind every divisibility rule you've ever memorized: it is just "what does 10 become, modulo this divisor?" Once you see that, you never have to memorize a rule again: you can derive it on the spot.

△ TRAPDOOR

Congruences are safe to add, subtract, and multiply — but *not* to divide carelessly. From $6 \equiv 0 \pmod 6$ you cannot cancel a 2 to get $3 \equiv 0$. Division (canceling a common factor) only works when that factor shares no common divisor with the modulus. When in doubt, don't divide — multiply by a proper inverse instead, and only when one exists.

CURIOUS CARL • TRAPDOOR!

Staring at $6 \equiv 0 \pmod 6$, Carl brightens: "Both sides share a 2 — cancel it! So $3 \equiv 0 \pmod 6$." He reaches to write it down.

The giant congruence, mid-relief at having shrunk, freezes: *"Hold on — 3 is nowhere near a multiple of 6! You may only cancel a factor that's coprime to the modulus."* In modular arithmetic, division is a privilege, not a right — and 2 never earned it mod 6.

The showpiece: a fact true for every integer at once

Now a claim that should look impossible to verify — it's about *all* whole numbers — yet modular arithmetic dispatches it cleanly.

WONDER THIS

Pick any whole number n at all. Compute $n^5 - n$.

I claim the result is *always* divisible by 30 — for every n, with no exceptions. Try $n = 2$: $32 - 2 = 30$. Try $n = 3$: $243 - 3 = 240 = 30 \times 8$. **Why must 30 always divide $n^5 - n$?**

The trick is to break the problem into its prime pieces. Since $30 = 2 \times 3 \times 5$, a number is divisible by 30 exactly when it's divisible by 2, by 3, *and* by 5 all at once. So we just need to check those three separately — and each is easy.

WATCH THIS — **#2 Splitting 30 into primes**

Claim. $30 \mid n^5 - n$ for every integer n.

First, factor: $n^5 - n = n(n^4 - 1) = n(n^2 - 1)(n^2 + 1) = (n-1)n(n+1)(n^2+1)$.

Divisible by 2. Among the consecutive integers $n-1, n, n+1$, at least one is even. So $2 \mid n^5 - n$.

Divisible by 3. Among three consecutive integers $n-1, n, n+1$, one is a multiple of 3. So $3 \mid n^5 - n$.

Divisible by 5. This is the pretty one. Work modulo 5: every integer is $\equiv 0, 1, 2, 3,$ or 4. Checking the fifth powers,

$$0^5 \equiv 0, \quad 1^5 \equiv 1, \quad 2^5 = 32 \equiv 2, \quad 3^5 = 243 \equiv 3, \quad 4^5 = 1024 \equiv 4 \pmod{5}.$$

In every case $n^5 \equiv n \pmod{5}$, so $5 \mid n^5 - n$. (This is a glimpse of *Fermat's Little Theorem*, the star of the next chapter.)

$2 \mid n^5 - n$ $\quad$ $3 \mid n^5 - n$ $\quad$ $5 \mid n^5 - n$ $\xrightarrow{\text{so}}$ $30 \mid n^5 - n$

Combine. Since 2, 3, and 5 each divide $n^5 - n$ and they share no common factors, their product 30 divides it too. ■

The strategy here is the one to carry away: *to prove divisibility by a composite number, split it into prime-power pieces and handle each separately.* A statement about all infinitely many integers reduced to checking five residues mod 5 and a parity-and-thirds observation — finite, fast, and complete. This split-into-primes principle is so useful it has a name when run in reverse, the *Chinese Remainder Theorem*, which says the pieces always reassemble uniquely. Picture it as *two clocks*: solving $x \equiv a \pmod{m}$ and $x \equiv b \pmod{n}$ is asking for a moment when a size-m clock reads a while a size-n clock reads b. When the two clock sizes are *coprime*, every pair of readings occurs exactly once before the combined clock (of size mn) repeats — so there is always a unique solution mod mn. One awkward modulus becomes two easy ones.

TRY TO BREAK IT

The digit-sum test for divisibility by 3 works only because $10 \equiv 1 \pmod{3}$, so every power of ten collapses to 1. Try to cook up the same tidy digit-sum rule for divisibility by 7: what is 10 mod 7, and then 100 mod 7, and 1000 mod 7? Watch the powers of ten *refuse* to settle on 1 (or even on ± 1), and you'll see exactly why 7 has no clean digit-sum rule — and why 9 and 11 are the lucky ones.

Payoff: taming an astronomical power

Here's the kind of problem modular arithmetic was born for.

WONDER THIS

What is the remainder when 2^{100} is divided by 7?

2^{100} is a thirty-one-digit number. **But we only want its remainder mod 7** — and for that, we never need to compute the monster itself.

The key is to find where the powers of 2 start *repeating* mod 7. Compute a few: $2^1 \equiv 2, 2^2 \equiv 4, 2^3 = 8 \equiv 1 \pmod{7}$. The moment we hit 1, the cycle resets — because multiplying by further factors of $2^3 \equiv 1$ changes nothing. So the powers of 2 cycle with period 3: $2, 4, 1, 2, 4, 1, \ldots$

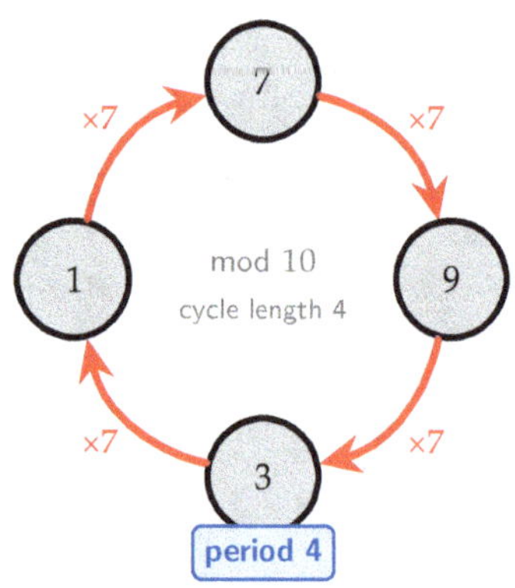

The last digit of 7^n tells the same story — it cycles $7, 9, 3, 1$ with period 4.

Now $2^{100} = 2^{3\times33+1} = (2^3)^{33} \cdot 2^1 \equiv 1^{33} \cdot 2 = 2 \pmod{7}$. **The remainder is** 2 — found with nothing but the observation that $2^3 \equiv 1$ and a quick division of the exponent by 3. The whole strategy is to *find the cycle length, then reduce the exponent modulo that length.* A thirty-one-digit computation became counting to three. *This is the entire reason anyone bothers with remainders: they turn the colossal into the trivial.*

THE TURN *from grind to insight*

The obvious plan is to actually build 2^{100} — double, and double, and double, a hundred times over — until you hold the thirty-one-digit monster, then divide the whole slab by 7 and keep the scraps. You'd be multiplying all day and risking a slip at every step, all to throw away every digit but one. **The creative turn** is to discard the giant before it ever forms: carry only the *remainder* mod 7 at each step, and watch the powers of 2 fall into the short loop $2, 4, 1, 2, 4, 1, \ldots$ The colossus collapses into a cycle of length three you can read off, so the exponent's only job is to tell you where in the loop you land. It wins because the size of the number was never the point — only its place on a tiny repeating clock.

The modulus that kills an equation

So far modular arithmetic has *computed* things — last digits, remainders, whether a number is divisible. Its most creative use runs the other way: it proves that something can *never* happen.

WONDER THIS

Find whole numbers x and y with $x^2 + y^2 = 2023$.

Go ahead and hunt: $44^2 = 1936$ leaves 87 (not a square); $43^2 = 1849$ leaves 174 (not a square); and on it goes. Every attempt fails — **but failing to *find* a solution is not a proof that none exists.** There are infinitely many pairs to try. How could you prove that *not one* of them works?

You don't search the infinite list. You find a single modulus under which the equation cannot even get off the ground.

THE MOVE

The impossibility filter. To prove an equation has *no* integer solutions, reduce both sides modulo a well-chosen m. Each side can only land on certain residues; if those two sets of residues *never overlap*, no solution can exist — one congruence kills infinitely many candidates at once. The whole art is choosing m so the pieces are tightly constrained. Two workhorses: *squares are only* 0 *or* 1 *mod* 4, and *cubes are only* 0 *or* ± 1 *mod* 9.

WATCH THIS **#3 Two equations, each killed in one line**

A sum of two squares, mod 4. Every square is 0 or 1 modulo 4 (even numbers square to 0, odd numbers to 1). Adding two of them,

$$x^2 + y^2 \equiv 0{+}0,\ 0{+}1,\ \text{or } 1{+}1 \equiv 0,\ 1,\ 2 \pmod 4.$$

A sum of two squares is therefore *never* $\equiv 3 \pmod 4$. But $2023 = 4\cdot 505 + 3 \equiv 3$, so $x^2 + y^2 = 2023$ has no integer solutions — settled without testing a single pair.

A sum of three cubes, mod 9. Running $0^3, 1^3, \ldots, 8^3$ through mod 9 leaves only 0, 1, and $8 \equiv -1$.

So each cube is 0 or ± 1, and three of them can only sum to

$$0,\ 1,\ 2,\ 3,\ 6,\ 7,\ 8 \pmod 9 \qquad \text{(that is, } 0, \pm 1, \pm 2, \pm 3\text{)},$$

which *skips* 4 *and* 5. Since $4 \equiv 4 \pmod 9$, the equation $x^3 + y^3 + z^3 = 4$ has no integer solutions either. ■

Notice what just happened. The conventional instinct (plug in numbers and hope to either land a solution or "see" why none works) can run forever, because the candidate list is infinite and a near-miss proves nothing. The switch is to stop searching the solutions and *search for the right modulus* instead: one lens under which the two sides live in disjoint worlds. Find it, and infinitely many candidates die in a single line. *To prove nothing works, don't test the candidates — change the lens until the impossibility is obvious.*

△ **TRAPDOOR**

The filter runs one way only. A well-chosen modulus can prove a solution is *impossible*, but passing the test proves nothing: if the residues *do* overlap, you have only learned that *this* modulus cannot rule the equation out — not that a solution exists. (Some equations are solvable modulo *every* m yet have no integer solution at all.) And a badly chosen modulus is simply silent: reduce $x^2 + y^2 = 2023$ mod 3 instead of mod 4 and the residues overlap, telling you nothing. The skill is picking the modulus that squeezes the pieces hardest — squares mod 4 (or mod 8), cubes mod 9, kth powers mod a prime where they collapse.

TRY TO BREAK IT

Push the squares one modulus further: check that modulo 8, every square is 0, 1, or 4. Now list what *three* such values can add up to mod 8 — and watch one residue, $7 \pmod 8$, stay forever out of reach. That single gap is the visible tip of Legendre's three-square theorem: a number is a sum of three squares *unless* it has the shape $4^a(8b + 7)$.

Choose the modulus like a lens. The whole art of the impossibility filter is picking the world where the obstruction becomes visible, and a few lenses do most of the work:

- *mod* 2 — parity (odd/even);
- *mod* 4 (or 8) — squares, which see only $\{0, 1\}$ mod 4 and $\{0, 1, 4\}$ mod 8;
- *mod* 9 — cubes, which see only $\{0, \pm 1\}$;
- *mod* p (prime) — inverses and quadratic residues.

When an equation mixes squares, cubes, or high powers, reach for the lens under which *that* kind of term is most sharply constrained.

Undoing multiplication: modular inverses

The trapdoor warned us never to divide a congruence carelessly. But sometimes we genuinely need to "undo" a multiplication — to solve $7x \equiv 1 \pmod{11}$, say. In ordinary arithmetic we'd divide by 7; mod 11 there are no fractions, so we need a substitute. That substitute is the *modular inverse*.

THE MOVE

The *inverse* of a modulo m is a number a^{-1} with $a \cdot a^{-1} \equiv 1 \pmod{m}$. It plays the role $1/a$ plays for ordinary numbers: multiplying by a^{-1} cancels a factor of a. The catch — and it is the whole story — is that a^{-1} *exists if and only if* $\gcd(a, m) = 1$. When a shares a factor with m, no multiple of a can ever land on 1, so there is nothing to undo with.

Why does coprimality decide everything? Because $ax \equiv 1 \pmod{m}$ means $ax - 1$ is a multiple of m, i.e. $ax + my = 1$ for some integer y — and that equation is solvable in integers exactly when $\gcd(a, m) = 1$ (the Extended Euclidean Algorithm produces the solution). So finding an inverse is the same problem as running Euclid's algorithm backward. For small moduli you can often skip the machinery and just hunt for the multiple of a that is one more than a multiple of m.

WATCH THIS — **#4 Finding a modular inverse**

Goal. Find $7^{-1} \pmod{11}$ — a number x with $7x \equiv 1 \pmod{11}$.

First, check it exists. $\gcd(7, 11) = 1$ (both are prime and distinct), so an inverse is guaranteed.

Trial method. Run through multiples of 7 and reduce mod 11, looking for 1:

$$7 \cdot 1 = 7, \quad 7 \cdot 2 = 14 \equiv 3, \quad 7 \cdot 3 = 21 \equiv 10, \quad \ldots, \quad 7 \cdot 8 = 56 = 55 + 1 \equiv 1 \pmod{11}.$$

There it is: $7 \cdot 8 \equiv 1 \pmod{11}$, so $7^{-1} \equiv 8 \pmod{11}$.

Using it. Now we can "divide by 7" mod 11. To solve $7x \equiv 5 \pmod{11}$, multiply both sides by 8: $x \equiv 8 \cdot 5 = 40 \equiv 7 \pmod{11}$. Check: $7 \cdot 7 = 49 = 44 + 5 \equiv 5 \pmod{11}$. ■

△ TRAPDOOR

Inverses are not free — they exist *only* when $\gcd(a, m) = 1$. There is no inverse of 6 modulo 9, because $\gcd(6, 9) = 3$: every multiple of 6 is $0, 6, 3, 0, 6, 3, \ldots \pmod{9}$, never 1. So $6x \equiv 1 \pmod{9}$ has *no* solution. Before you try to cancel a factor in a congruence, check that it is coprime to the modulus; if it isn't, the cancellation is illegal.

Which numbers are squares: quadratic residues

Mod 7, is there an x with $x^2 \equiv 3$? With $x^2 \equiv 2$? One of these has a solution and the other doesn't, and the pattern of *which numbers are perfect squares modulo p* is one of the oldest and richest questions in number theory.

THE MOVE

A nonzero residue r is a *quadratic residue* (QR) mod p if $x^2 \equiv r \pmod{p}$ has a solution — i.e. r is a perfect square mod p. Otherwise it's a *non-residue*. To find them all, you don't test every r: you just *square every residue* and see what comes out. And because x and $-x$ square to the same thing, the squares pair up — so exactly $(p-1)/2$ of the nonzero residues are QRs, and the other $(p-1)/2$ are not.

WATCH THIS — *#5 Which numbers are squares mod 7*

Goal. Find every nonzero quadratic residue mod 7.

Square each nonzero residue and reduce:

$$1^2 \equiv 1, \quad 2^2 \equiv 4, \quad 3^2 = 9 \equiv 2, \quad 4^2 = 16 \equiv 2, \quad 5^2 = 25 \equiv 4, \quad 6^2 = 36 \equiv 1 \pmod{7}.$$

Notice the symmetry: the list reads $1,4,2,2,4,1$ — a palindrome, because $4 \equiv -3$, $5 \equiv -2$, $6 \equiv -1$ square to the same values as $3,2,1$. The distinct outputs are

$$\{1,2,4\} \ \text{(quadratic residues)}, \qquad \{3,5,6\} \ \text{(non-residues)}.$$

So $x^2 \equiv 2 \pmod{7}$ *is* solvable ($x \equiv 3$ or 4), but $x^2 \equiv 3 \pmod{7}$ is *not*. There are exactly $(7-1)/2 = 3$ of each — precisely the split the move predicted. ■

Trigger A question asking whether something is a perfect square mod p, or to find/solve $x^2 \equiv r$, or to count squares.

Do first Square every residue $1^2, 2^2, \ldots, (\frac{p-1}{2})^2$ and reduce — those outputs are exactly the QRs (there are $(p-1)/2$ of them).

Trapdoor Forgetting the residues pair as x and $-x$, so you only need to square up to $(p-1)/2$; and that a non-residue is never a square no matter what.

Upgrade The deeper laws — the product of two non-residues is a residue, and -1 is a QR exactly when $p \equiv 1 \pmod{4}$ — open the door to quadratic reciprocity.

SEEN BEFORE?

Quadratic residues quietly use the modular inverses we just built: when p is prime, every nonzero residue is coprime to p, so *every* nonzero residue has an inverse, and the squares form exactly half of them. The two ideas — inverses exist (coprimality) and squares are half (pairing) — are the first facts about the arithmetic *inside* a prime modulus.

GOING FURTHER — *optional • beyond AIME*

This box is an optional detour, just past contest level; the listing technique above is all you need for the problems. But once you can list quadratic residues, it is natural to ask *which* ones appear without grinding out the whole table — and there is a beautiful test.

Euler's criterion. For an odd prime p and a nonzero residue a,

$$a \text{ is a quadratic residue mod } p \iff a^{(p-1)/2} \equiv 1 \pmod{p},$$

and a is a non-residue exactly when $a^{(p-1)/2} \equiv -1 \pmod{p}$. (By Fermat's Little Theorem the square of $a^{(p-1)/2}$ is $a^{p-1} \equiv 1$; and modulo a prime, $x^2 \equiv 1$ forces $(x-1)(x+1) \equiv 0$, hence $x \equiv \pm 1$ — so the power can only be ± 1.) This converts "is a a square?" into a single exponentiation.
When is -1 a square? Feed $a = -1$ into the criterion:

$$(-1)^{(p-1)/2} \equiv 1 \pmod{p} \iff \tfrac{p-1}{2} \text{ is even} \iff p \equiv 1 \pmod{4}.$$

So -1 is a quadratic residue mod p *exactly* when $p \equiv 1 \pmod{4}$. Check it against our small cases: mod 5 (and $5 \equiv 1 \pmod{4}$) we have $2^2 = 4 \equiv -1$, so -1 is a square; mod 7 (and $7 \equiv 3 \pmod{4}$) the residues were $\{1, 2, 4\}$ and $-1 \equiv 6$ is missing, so -1 is a non-residue. This single fact is the first step on the road to *quadratic reciprocity*, one of the jewels of number theory.

What to try when you're stuck

When a problem involves divisibility, remainders, last digits, or very large powers:

1. **Switch to remainders.** Replace every number by its remainder mod m as early as possible, keeping arithmetic small.
2. **For divisibility rules,** ask what the base (10) reduces to mod the divisor: 1 gives digit sums, -1 gives alternating sums.
3. **For composite divisors,** split into prime-power factors and prove divisibility by each separately, then recombine.
4. **For large powers,** find the cycle: compute powers mod m until they repeat, then reduce the exponent modulo the cycle length.
5. **To prove *no* solutions exist,** reduce the equation mod a sharp modulus (squares mod 4, cubes mod 9); if the two sides can never share a residue, no solution can exist.
6. **Factor when you can.** Algebraic factorizations like $n^5 - n = (n-1)n(n+1)(n^2+1)$ expose consecutive-integer divisibility.
7. **To divide, invert.** To solve $ax \equiv b \pmod{m}$ when $\gcd(a, m) = 1$, find a^{-1} (trial or Extended Euclid) and multiply: $x \equiv a^{-1}b$.
8. **For "is it a square?",** square every residue up to $(m-1)/2$; the outputs are exactly the quadratic residues.

THE MOVE

One-page recap. Modular arithmetic replaces numbers by remainders, which add, subtract, and multiply like ordinary numbers. Toolbox: **factor the base** (divisibility rules from $10 \equiv 1$ or -1; the $1001 = 7 \cdot 11 \cdot 13$ trick), **split a composite modulus into primes** ($30 = 2 \cdot 3 \cdot 5$, then recombine via the Chinese Remainder Theorem), and **find the cycle of powers** ($2^3 \equiv 1 \pmod 7$, so reduce the exponent mod 3). To undo a multiplication, **multiply by an inverse** (which exists exactly when $\gcd(a, m) = 1$); to ask whether something is a square, **list the quadratic residues** (square every residue — exactly half are QRs); and to prove an equation has *no* solutions, **reduce mod a sharp modulus** (squares are $0, 1 \bmod 4$; cubes $0, \pm 1 \bmod 9$) until the two sides cannot match. Don't divide congruences carelessly. When numbers get huge, care only about their remainders.

THE NAME FOR IT

Modular arithmetic

Working with remainders as if they were numbers is *modular arithmetic*, written $a \equiv b \pmod m$ for *congruence*. The principle that lets a composite modulus split into its prime parts is the *Chinese Remainder Theorem*.

WHERE IT LEADS Modular arithmetic is the first *quotient ring* $\mathbb{Z}/m\mathbb{Z}$ — the gateway to *abstract algebra*. Congruences, the Chinese Remainder Theorem, and finite fields underpin all of modern *cryptography* (RSA, elliptic curves), *coding theory*, and algebraic number theory.

CURIOUS CARL • EUREKA

Carl watches 7^{100} mop its brow and shrink to a single digit on the clock face. "I was going to write you out in *full*. On the long paper. And you'd have folded down to *this* the whole time."

SAME BEAST, NEW COSTUME

The same "only the remainder matters" beast, in disguise across the book:

Last-digit cycles → Divisibility rules → Impossibility filters → Chinese Remainder → Fermat/Euler → Roots-of-unity filter.

Each one throws the giant away and keeps only what a small modulus can see. Find the right modulus, and the colossal question shrinks to arithmetic you can do in your head.

The practice ladder

THE PRACTICE LADDER *climb at your own pace*

LAUNCH *Warm the engine. Nudges below.*

1. What is the remainder when 1234 is divided by 9? Use the digit-sum rule.

2. Show that $\overline{ababab}$ (a two-digit block repeated three times) is always divisible by 10101, and factor 10101.

3. Find the inverse of 5 modulo 13 — a number x with $5x \equiv 1 \pmod{13}$ — by trying multiples of 5.

CORE BUILDER *Master the central moves.*

1. Find the last digit of 7^{2024} by finding the cycle of 7^k modulo 10.

2. Prove that $n^3 - n$ is divisible by 6 for every integer n.

3. Derive the divisibility-by-11 rule by reducing each power of 10 modulo 11.

4. Solve $3x \equiv 4 \pmod{7}$ by finding the inverse of 3 mod 7 and multiplying.

STRETCH *Transfer to unfamiliar ground.*

1. Find the remainder when 3^{100} is divided by 100 (find the cycle of 3^k mod 100).

2. Solve the simultaneous congruences $x \equiv 2 \pmod{3}$ and $x \equiv 3 \pmod{5}$, and find all solutions mod 15.

3. List all the nonzero quadratic residues mod 11 by squaring each residue, and check that there are exactly $(11-1)/2 = 5$ of them.

THE FLASH *No clock — each cracks open in one flash of insight, once you find the angle. Nudges at the back.*

1. What is 5^{99} mod 7?

2. Is 918082 divisible by 11?

PROOF LAB *Write it clean, in full sentences.*

1. Prove that for every integer n, $30 \mid n^5 - n$, by splitting 30 into primes and handling each.

2. Prove that $\overline{abcabc} = \overline{abc} \times 1001$ and conclude divisibility by 7, 11, and 13; then state and prove the analogous fact for a four-digit block repeated twice.

3. Prove that no perfect square ends in the digit 2, 3, 7, or 8. (*Work mod* 10.)

4. Show that $x^2 - 5y^2 = 2$ has no integer solutions. (*Reduce mod* 5.)

5. Prove that $x^3 + y^3 = 3$ has no integer solutions. (*Reduce mod* 9.)

REFLECTION *No nudge. Write a few sentences.*

1. For the large-power problems, how did finding the cycle length change the difficulty? Describe the moment the "astronomical" computation became small.

NUDGES

Launch. (1) Digit sum $10 \equiv 1 \pmod 9$, so remainder 1. (2) $\overline{ababab} = \overline{ab} \times 10101$, and $10101 = 3 \times 7 \times 13 \times 37$. (3) $5 \cdot 8 = 40 = 39 + 1 \equiv 1 \pmod{13}$, so $5^{-1} \equiv 8$. **Core.** (4) 7^k mod 10 cycles 7, 9, 3, 1 with period 4; $2024 \equiv 0$, so last digit 1. (5) $n^3 - n = (n-1)n(n+1)$, three consecutive integers, divisible by 2 and 3. (6) $10 \equiv -1 \pmod{11}$, so $10^k \equiv (-1)^k$, giving the alternating sum. (7) $3^{-1} \equiv 5 \pmod 7$ (since $3 \cdot 5 = 15 \equiv 1$); $x \equiv 5 \cdot 4 = 20 \equiv 6 \pmod 7$. **Stretch.** (8) 3^k mod 100 cycles with period 20; $3^{100} = (3^{20})^5 \equiv 1$, remainder 01. (9) $x \equiv 8 \pmod{15}$. (10) Squares mod 11: 1, 4, 9, 5, 3, 3, 5, 9, 4, 1, so the QRs are $\{1, 3, 4, 5, 9\}$—exactly 5. **The Flash.** (11) $5 \equiv -2 \pmod 7$ and $5^3 \equiv -1$, so $5^{99} = (5^3)^{33} \equiv (-1)^{33} \equiv -1 \equiv 6 \pmod 7$. (12) Alternating sum $2-8+0-8+1-9 = -22$, divisible by 11; yes. **Proof.** (13) Factor $n^5 - n = n(n^2-1)(n^2+1) = (n-1)n(n+1)(n^2+1)$. Among the consecutive integers $n-1, n, n+1$ one is even and one is a multiple of 3, so 2 and 3 both divide the product; by Fermat $n^5 \equiv n \pmod 5$, so $5 \mid n^5 - n$ as well. Since 2, 3, 5 are pairwise coprime, their product 30 divides $n^5 - n$. (14) $\overline{abcabc} = \overline{abc} \cdot 1000 + \overline{abc} = \overline{abc}(1001)$, and $1001 = 7 \times 11 \times 13$, so 7, 11, 13 each divide it. A four-digit block repeated gives $\overline{wxyzwxyz} = \overline{wxyz} \cdot 10001$ with $10001 = 73 \times 137$, hence divisible by 73 and 137. (15) The last digit of n^2 depends only on the last digit of n; squaring 0–9 gives last digits 0, 1, 4, 9, 6, 5, 6, 9, 4, 1, so a square ends only in 0, 1, 4, 5, 6, 9 — never 2, 3, 7, 8. (16) Reduce mod 5: since $5y^2 \equiv 0$, the equation forces $x^2 \equiv 2 \pmod 5$. But squares mod 5 are only 0, 1, 4, so this is impossible and there are no integer solutions. (17) Cubes mod 9 are only $0, \pm 1$ (check 0–8), so $x^3 + y^3 \in \{0, \pm 1, \pm 2\} = \{0, 1, 2, 7, 8\} \pmod 9$, which omits 3; hence no integer solutions.

Checkpoint: nine ways to escape

Nine chapters in, pause and look back at what they share. Each one drilled the same reflex in a new disguise: when the *ordinary* method — listing, searching, simulating, expanding, grinding — explodes or buries the structure, there is a *creative escape* that steps off the usual battlefield entirely. Here they are, side by side. When a problem has you reaching for the trap in the middle column, let your eye slide to the right: the escape is already waiting.

When you're stuck with...	**the ordinary trap**	**the creative escape**	**Ch.**
a nasty count	list every case	count a *different* set	1
forced repetition	hunt for the pair	compress into too-few boxes	2
a moving puzzle	simulate the moves	find what *can't* change	3
a hard object to count	count it head-on	encode it as an easy one	4
a "size n" problem	attack the whole thing	solve a smaller shadow	5
a long, ugly sum	add it term by term	make the middle cancel	6
a coefficient identity	expand blindly	read it as a story	7
an infinite sequence	compute terms forever	hang it on a polynomial	8
a giant number	compute the monster	keep only its remainder	9

Read across any row and you feel the book's one promise: *don't fight the problem on its own terms — change the terms.* Read *down* the last column and you are holding the first half of a creative solver's toolkit — nine doors the problem never knew it left open.

The chapters ahead open more of them. Powers will collapse into cycles, inequalities will melt into balance, real problems will detour through the imaginary plane, geometry will flatten under affine maps, symmetry will average itself away, networks will shrink to local counts, and whole distributions will dissolve into tiny averages. Different doors, same instinct: when the direct road is blocked, look for the one the problem forgot to lock.

Chapter Ten

10

When Giants Collapse

Primes are the atoms of arithmetic — and they obey hidden laws that make impossible-looking powers collapse to a single digit.

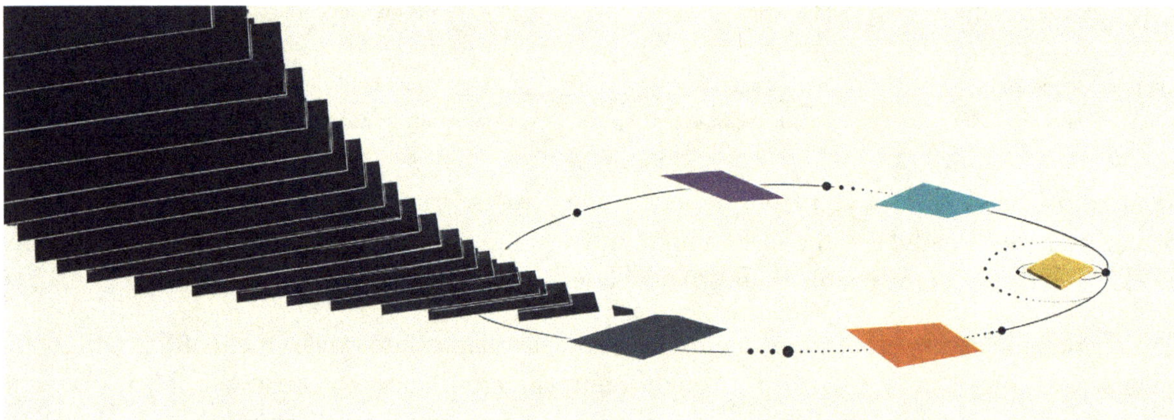

Fermat & Euler: big exponents collapse.

Let's raise the stakes from last chapter with a genuinely intimidating number.

WONDER THIS

What are the last two digits of 7^{2026}?

That's 7 raised to the two-thousand-twenty-sixth power — a number with over seventeen hundred digits. **Find its final two digits.**

In Chapter 9 we tamed 2^{100} mod 7 by hunting for the cycle of powers. We could do that here too. But searching for the cycle of 7^k mod 100 by brute force would take a while. This chapter hands us a theorem that tells us the cycle length *in advance*, without any searching. It's one of the most useful facts in number theory, and it's named for Fermat.

CURIOUS CARL • FIRST GUESS

Carl: "For 3^{100} mod 7? A hundred multiplications by 3, then one division at the very end." (*He starts stacking threes.*)

The catch: Fermat says $3^6 \equiv 1 \pmod 7$, Carl — so the exponent wraps around mod 6, and a hundred multiplications shrink to a handful.

The set-up: "last two digits" means working modulo 100. We need to know how the powers of 7 behave mod 100, and specifically when they return to 1. Fermat's idea, generalized by Euler, says: count the numbers below 100 that share no common factor with 100 (call that count $\varphi(100)$), and *that* is a cycle length. It turns out $\varphi(100) = 40$, which means

$$7^{40} \equiv 1 \pmod{100}.$$

Once we know that, the giant exponent crumbles. Divide 2026 by 40: it goes 50 times with remainder 26. So $7^{2026} = (7^{40})^{50} \cdot 7^{26} \equiv 1^{50} \cdot 7^{26} = 7^{26} \pmod{100}$. A short computation of 7^{26} mod 100 (squaring repeatedly, staying mod 100) gives 49. **The last two digits of 7^{2026} are 49**, and the seventeen-hundred-digit monster never had to be written down.

THE TURN *from grind to insight*

Last chapter's method would have us chase the powers of 7 mod 100 one multiplication at a time, hunting blindly for the moment they loop back to 1 — and with a cycle of length 40 hiding in there, that's a long, patient slog before the giant exponent 2026 even becomes attackable. **The decisive shortcut** is to stop searching: Fermat and Euler hand you the cycle length *in advance*, telling you $7^{40} \equiv 1 \pmod{100}$ before you multiply a single thing. The seventeen-hundred-digit tower then collapses in one stroke — reduce 2026 mod 40 down to 26, and you're left with a tiny remainder to finish. It wins because you no longer discover the rhythm by hand; the theorem names the tempo, and the colossus falls in a line.

THE MOVE

Fermat's Little Theorem. If p is prime and a is not a multiple of p, then

$$a^{p-1} \equiv 1 \pmod{p}.$$

Euler's generalization. For any modulus m, if a shares no common factor with m, then $a^{\varphi(m)} \equiv 1 \pmod{m}$, where $\varphi(m)$ counts the integers in $\{1, \dots, m\}$ coprime to m. These theorems hand you a cycle length for free — so any power a^N reduces by taking N modulo that cycle length.

This is the payoff of thinking in remainders: the primes impose a rhythm on the powers, and the theorem tells you the tempo without making you find it by hand. Whole families of "find the last digits" and "find the remainder" problems fall to a single division of the exponent.

The hidden clock. A modulus turns powers into a *clock*: multiply by a and the hand steps forward one notch, and after finitely many steps it must return to where it started. Fermat and Euler promise the clock *resets*, and hand you a guaranteed reset time, $\varphi(m)$; the **order** (next section) is the true, smallest reset time. So the real question is never "how big is a^N?" but "*where does N land on the clock?*", which depends only on N modulo the reset time. *Don't compute the power; find its clock.* (And a big clock often splits: to read 7^{2026} mod 100, run it on a mod-4 clock and a mod-25 clock and recombine, exactly the two-clocks trick of Chapter 9.)

Orders: the true cycle length

Euler's theorem gives *a* cycle length, $\varphi(m)$ — but sometimes the powers return to 1 even sooner. The *actual* smallest cycle length is called the *order* of a, and understanding it sharpens every computation.

WATCH THIS **#1 The order divides the rest**

Claim. Work modulo 7. The powers of 2 are $2, 4, 1, 2, 4, 1, \ldots$ — so the order of 2 is 3. Fermat guarantees $2^6 \equiv 1$; notice 3 divides 6. This is no accident.

The key principle. If $a^d \equiv 1 \pmod{m}$ and the *order* of a is t (the smallest positive power giving 1), then t must divide d.

Why. Divide d by t: write $d = qt + r$ with $0 \le r < t$. Then

$$1 \equiv a^d = (a^t)^q \cdot a^r \equiv 1^q \cdot a^r = a^r \pmod{m}.$$

So $a^r \equiv 1$ with r smaller than the order t. But t was the *smallest* positive power giving 1 — the only way out is $r = 0$. Hence $t \mid d$. ∎

Trigger Any question about cycles of powers, "when does a^k first return to 1," or relating different exponents that give 1.

Do first Find the order t (smallest power giving 1); then every exponent giving 1 is a multiple of t, and t divides $\varphi(m)$.

Trapdoor Assuming the order equals $\varphi(m)$. It only *divides* $\varphi(m)$ — it's often strictly smaller, so check.

Upgrade Because the order divides $\varphi(m)$, you only need to test divisors of $\varphi(m)$ as candidate cycle lengths — a huge shortcut.

This divisibility is enormously practical: since the order always divides $\varphi(m)$, you never test every exponent, only the divisors of $\varphi(m)$. The order is the heartbeat of a modulo m, and everything about a's powers (when they repeat, which residues they hit) is governed by that single number. Recognizing that the order *divides* the obvious cycle length is what turns a search into a quick check.

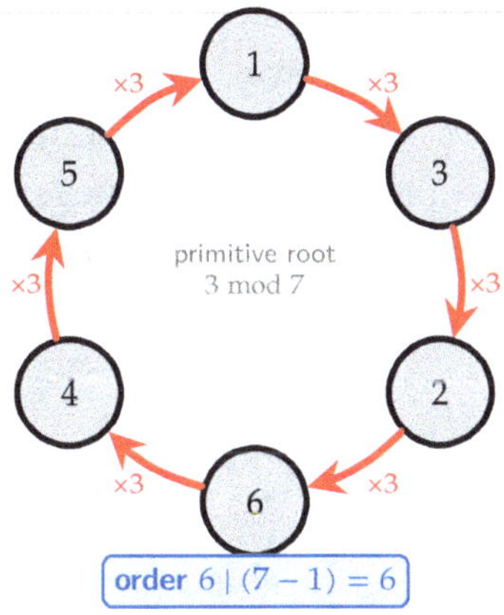

The primitive root 3 modulo 7 sweeps all six nonzero residues; its order 6 divides 7 − 1.

△ TRAPDOOR

Fermat's Little Theorem requires a *not* divisible by p (and Euler's requires a coprime to m). If a shares a factor with the modulus, the theorem simply doesn't apply — $a^{\varphi(m)}$ need not be 1. Always check the coprimality condition before invoking the theorem; forgetting it is the most common Fermat error.

CURIOUS CARL • TRAPDOOR!

Carl fires off the theorem on autopilot: "$a^{p-1} \equiv 1$, so $7^6 \equiv 1 \pmod 7$!" He's already moving to the next line.

The narrator catches his sleeve: *Careful, Carl — here* $p = 7$ divides $a = 7$, *so* $\gcd(a,p) \neq 1$ *and Fermat says nothing; in fact* $7^6 \equiv 0 \pmod 7$, *not* 1. The coprimality clause isn't fine print — it's the whole hinge the theorem swings on.

GOING FURTHER *optional • beyond AIME*

This is an optional, beyond-contest detour — skip it on a first pass and lose nothing essential for the problems ahead. Sometimes a single element's powers don't merely cycle but sweep through *every* nonzero residue before returning to 1. Recall that the *order* of a modulo p is the smallest positive t with $a^t \equiv 1$, and Fermat forces the order to divide $p-1$, so the largest it could possibly be is $p-1$ itself. When the order *hits* that maximum, a is called a *primitive root* modulo p: its powers $a^1, a^2, \dots, a^{p-1}$ run through all $p-1$ nonzero residues with no repeats before looping. For every prime p at least one primitive root exists — a single generator from which the entire multiplicative world mod p is built, turning "multiply mod p" into "add the exponents."
To see one, test 3 modulo 7 by listing its powers, always reducing mod 7:

$$3^1 \equiv 3, \quad 3^2 \equiv 2, \quad 3^3 \equiv 6, \quad 3^4 \equiv 4, \quad 3^5 \equiv 5, \quad 3^6 \equiv 1.$$

The six powers hit $3,2,6,4,5,1$ — *every* nonzero residue $\{1,2,3,4,5,6\}$ exactly once before the cycle closes at 1. So the order of 3 is $6 = \varphi(7) = 7-1$, and 3 is a primitive root mod 7. There is no formula that hands you a primitive root: you hunt for one, testing small candidates $2, 3, 5, \ldots$ and checking that no proper divisor of $p-1$ already returns the element to 1. The lasting idea is simply that one always exists — the nonzero residues mod a prime form a single cycle, and a primitive root is the element that walks the whole loop.

The showpiece: counting the coprimes

Euler's theorem needs $\varphi(m)$, so we'd better be able to compute it. The *Euler totient* $\varphi(m)$ counts how many integers from 1 to m are coprime to m. Computing it by listing would be slow. But there's a beautiful formula, and deriving it shows the prime-splitting philosophy of the last chapter in full flower.

WONDER THIS

How many integers from 1 to 100 share *no common factor* with 100?

Listing and checking all 100 would be tedious. **Is there a structural way to count them?**

WATCH THIS — **#2 The totient product formula**

Claim. If $m = p_1^{a_1} p_2^{a_2} \cdots p_r^{a_r}$ in primes, then

$$\varphi(m) = m\left(1 - \frac{1}{p_1}\right)\left(1 - \frac{1}{p_2}\right)\cdots\left(1 - \frac{1}{p_r}\right).$$

The idea: remove each prime's multiples. Start with all m numbers. To be coprime to m, a number must avoid being divisible by any prime factor of m. Take $100 = 2^2 \cdot 5^2$, whose prime factors are just 2 and 5.

A fraction $\frac{1}{2}$ of the numbers from 1 to 100 are even — knock those out, keeping the fraction $\left(1 - \frac{1}{2}\right)$. Of what remains, a fraction $\frac{1}{5}$ are multiples of 5 — knock those out too, keeping $\left(1 - \frac{1}{5}\right)$. The crucial fact (a consequence of the Chinese Remainder Theorem from last chapter) is that "divisible by 2" and "divisible by 5" behave *independently*, so the surviving fractions simply multiply:

$$\varphi(100) = 100\left(1 - \frac{1}{2}\right)\left(1 - \frac{1}{5}\right) = 100 \cdot \frac{1}{2} \cdot \frac{4}{5} = 40.$$

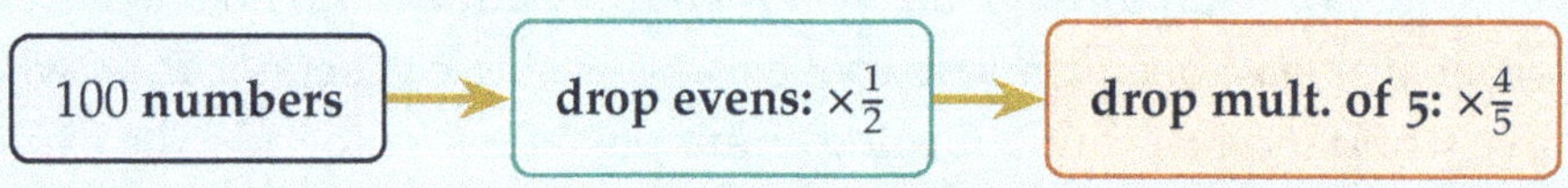

$100 \cdot \frac{1}{2} \cdot \frac{4}{5} = 40$

So exactly 40 integers from 1 to 100 are coprime to 100 — and that 40 is precisely the cycle length we used to crack 7^{2026}. ■

The formula is gorgeous because it's *multiplicative*. Each prime contributes its own independent factor, and the totient assembles from the primes exactly the way m itself does. This is the same lesson as the divisibility-by-30 proof: *the primes are the atoms, and arithmetic facts about m are built by combining facts about each prime separately.* Once you can factor m, the totient (and hence Euler's cycle length) costs you nothing.

SEEN BEFORE?

The independence of "divisible by 2" and "divisible by 5" is the *Chinese Remainder Theorem* foreshadowed in **Chapter 9**, and the $n^5 - n$ proof there was secretly Fermat's Little Theorem at $p = 5$. The order-divides-φ argument is the same "smallest positive thing" descent that powered the monovariants of **Chapter 3**.

TRY TO BREAK IT

Fermat's promise $a^{p-1} \equiv 1 \pmod{p}$ leans entirely on $\gcd(a, p) = 1$. Deliberately break that hinge: take $a = 7$ and $p = 7$, where p *does* divide a, and compute $7^6 \bmod 7$ by hand. Where does the "$\equiv 1$" land instead — and can you see why *no* multiple of p can ever satisfy the theorem? Watching the promise fail on purpose is the surest way to remember it always carries that condition.

Payoff: Wilson's elegant test

To close, a theorem of startling elegance: a property so clean it characterizes the primes exactly.

WONDER THIS

Take a number p, multiply together all the numbers below it — that's $(p-1)!$ — and look at the remainder mod p.

For $p = 7$: $6! = 720$, and $720 = 102 \times 7 + 6$, so $6! \equiv 6 \equiv -1 \pmod{7}$. **Is it always the case that $(p-1)! \equiv -1$ when p is prime?**

This is **Wilson's Theorem**: $(p-1)! \equiv -1 \pmod{p}$ if and only if p is prime. The idea behind it is lovely. Among the numbers $1, 2, \ldots, p-1$, work modulo a prime p: each one has a unique "partner" it multiplies with to give 1 (its inverse). Most numbers pair up with a *different* partner, and each such pair contributes a factor of 1 to the product. The only numbers that are their own partner are those satisfying $a^2 \equiv 1$ — which, modulo a prime, are exactly $a \equiv 1$ and $a \equiv -1$. So in the giant product $(p-1)!$, everything cancels in pairs except the lonely 1 and -1, leaving

$$(p-1)! \equiv 1 \cdot (-1) = -1 \pmod{p}.$$

All the middle terms annihilate in inverse-pairs, and only 1 and -1 survive. *It is telescoping's number-theoretic cousin — a great cancellation, leaving just the two endpoints.*

And the congruence *fails* for every composite number, which is what makes it a perfect (if impractical) primality test. The converse is quick: if $n = 4$, then $3! = 6 \equiv 2 \pmod{4}$, not -1; and if $n > 4$ is composite, its factors all sit among $1, 2, \dots, n-1$, so $n \mid (n-1)!$, giving $(n-1)! \equiv 0 \pmod{n}$, again not -1. So Wilson's congruence says *yes* to primes and *no* to everything else.

GOING FURTHER *optional • beyond AIME*

Another optional detour for the curious — safe to skip on a first pass. Fermat and Euler tell you *when* a power returns to 1; a companion tool answers a sharper question: not whether a prime divides $a^n - b^n$, but *exactly how many times* it does. Write $v_p(N)$ for the *p-adic valuation* of N — the largest k with $p^k \mid N$ — so $v_3(45) = 2$ since $45 = 3^2 \cdot 5$, and $v_5(100) = 2$ since $100 = 2^2 \cdot 5^2$. The *Lifting-the-Exponent (LTE) lemma* says: let p be an *odd* prime and a, b integers with $p \mid a - b$ but $p \nmid a$ and $p \nmid b$; then for every positive integer n,

$$v_p(a^n - b^n) = v_p(a - b) + v_p(n).$$

The power of p dividing $a^n - b^n$ is the power already in $a - b$, *plus* the power of p inside the exponent n itself.

For a worked example, take the largest power of 3 dividing $7^{100} - 1$ — a number with over eighty digits. Write it as $7^{100} - 1^{100}$, so $a = 7, b = 1, n = 100, p = 3$. The prime 3 is odd; $3 \mid a - b = 6$; and $3 \nmid 7, 3 \nmid 1$, so the lemma applies:

$$v_3(7^{100} - 1) = v_3(7 - 1) + v_3(100) = v_3(6) + v_3(100) = 1 + 0 = 1,$$

since $6 = 2 \cdot 3$ carries one factor of 3 while $100 = 2^2 \cdot 5^2$ carries none. So 3^1 exactly divides $7^{100} - 1$ (and 9 does not) — read off in one line. A word of caution: the hypotheses are not decoration. The prime must be *odd* (for $p = 2$ the formula changes), and you must have $p \mid a - b$ to start, or the lemma says nothing. For $a^n + b^n$ with n odd, apply the same lemma to $a^n - (-b)^n$, handling sums as well as differences.

What to try when you're stuck

When a problem involves primes, large powers modulo something, or the structure of remainders:

1. **Check coprimality**, then apply Fermat ($a^{p-1} \equiv 1 \bmod p$) or Euler ($a^{\varphi(m)} \equiv 1 \bmod m$) to get a cycle length for free.

2. **Reduce the exponent** modulo that cycle length — the giant power collapses to a small one.

3. **Find the order** (smallest power giving 1) for the sharpest cycle; remember it divides $\varphi(m)$, so only test divisors.

4. **Compute** $\varphi(m)$ by factoring m and applying the product formula $m \prod (1 - 1/p)$.

5. **Look for pairing/cancellation** in factorial products (Wilson) — inverses pair off, leaving only self-inverse terms.

(Two further tools — *primitive roots* and *Lifting the Exponent* — live in the optional "Going Further" boxes above, for when you want to reach past the contest-standard core.)

THE MOVE

One-page recap. Primes impose rhythm on powers. Toolbox: **Fermat** ($a^{p-1} \equiv 1 \bmod p$) and **Euler** ($a^{\varphi(m)} \equiv 1 \bmod m$) give cycle lengths, so reduce exponents modulo them; **order** is the true cycle length and divides $\varphi(m)$; **totient product formula** $\varphi(m) = m\prod_{p|m}(1 - \frac{1}{p})$ computes the cycle from the prime factorization; and **Wilson** $(p-1)! \equiv -1 \bmod p$ via inverse-pairing characterizes primes. When a power is astronomical, let Fermat hand you its tempo. (For more: *primitive roots* and *LTE* wait in the optional Going Further boxes.)

THE NAME FOR IT

Fermat, Euler, order, and Wilson

The fact $a^{p-1} \equiv 1 \pmod p$ for prime p is *Fermat's Little Theorem*, generalized by *Euler's theorem* $a^{\varphi(m)} \equiv 1 \pmod m$, where φ is the *Euler totient*. The smallest power returning to 1 is the *multiplicative order*, and $(p-1)! \equiv -1$ is *Wilson's theorem*. (An element of order $p-1$ is a *primitive root*, and $v_p(a^n - b^n) = v_p(a-b) + v_p(n)$ is the *Lifting-the-Exponent lemma* — both optional extensions covered above.)

WHERE IT LEADS Orders and Fermat–Euler open the theory of *cyclic groups* and the *multiplicative group* of a finite field — core *group theory*. These ideas mature into *class field theory* and power the primality tests on which public-key cryptography rests. The optional *primitive root* (a *generator* of that cyclic group) and *Lifting-the-Exponent* (p-adic valuation) tools carry you further into olympiad territory.

A closing thought before you climb. The deep idea of this chapter is that exponentiation modulo m is not the wild, runaway operation it first appears to be. It is a clock. Multiply by a over and over and you do not wander off to infinity; you march around a finite circle of remainders and return, on a perfectly predictable beat, to where you began. Fermat and Euler tell you the length of that circle before you take a single step; the order tells you the shortest lap; Wilson reads the primes off the rhythm of how the remainders pair up. Once you start *hearing* the periodicity — once "7^{100} mod 13" sounds less like a calculation and more like "where does the hundredth tick land on a thirteen-hour clock?" — the astronomical exponents in the problems below stop being intimidating. They are just large numbers of laps around a small track, and you already know how to count laps. Keep the clock in mind as you work, and reach for the cycle length first, every time.

CURIOUS CARL • EUREKA

Carl glares at the tower of 3's he stacked to the ceiling. "You were only ever going to land on the same handful of remainders — and you let me build you all the way up first. Cruel."

SAME BEAST, NEW COSTUME

The same "exponents wrap around a cycle" beast, in disguise across the book:

Fermat's little theorem → Euler's theorem → order divides $\varphi(n)$ → Wilson's theorem.

Each one names the length of a hidden loop, so a colossal exponent only has to say where on the loop it lands. Find the cycle, reduce the exponent, and the giant collapses.

The practice ladder

THE PRACTICE LADDER *climb at your own pace*

LAUNCH *Warm the engine. Nudges below.*

1. Compute $\varphi(12)$, $\varphi(13)$, and $\varphi(36)$ using the product formula.
2. Use Fermat's Little Theorem to find 3^6 mod 7 without multiplying it out.
3. Find the order of 2 modulo 7 by listing its powers until they return to 1.

CORE BUILDER *Master the central moves.*

1. Find 3^{100} mod 7 by reducing the exponent with Fermat.
2. Find the order of 3 modulo 7, and verify it divides $\varphi(7) = 6$.
3. Compute the last two digits of 3^{2024} using Euler's theorem mod 100.

STRETCH *Transfer to unfamiliar ground.*

1. Prove that if p is prime, then $\binom{p}{k} \equiv 0 \pmod{p}$ for $0 < k < p$, and use it to give a slick proof of Fermat's Little Theorem by expanding $(1 + 1 + \cdots + 1)^p$.
2. Show that $a^{561} \equiv a \pmod{561}$ for all a even though $561 = 3 \times 11 \times 17$ is *not* prime (a Carmichael number), using the Chinese Remainder Theorem and Fermat on each prime.

THE FLASH *No clock — each cracks open in one flash of insight, once you find the angle. Nudges at the back.*

1. What is 7^{2026} mod 100?
2. Compute 6! mod 7 and 10! mod 11 using Wilson's theorem.

PROOF LAB *Write it clean, in full sentences.*

1. Prove the totient product formula $\varphi(m) = m\prod_{p|m}\left(1-\frac{1}{p}\right)$, justifying why the prime-factor conditions are independent.
2. Prove Wilson's theorem: for prime p, $(p-1)! \equiv -1 \pmod{p}$, by pairing each residue with its inverse and identifying the self-inverse elements.

REFLECTION *No nudge. Write a few sentences.*

1. How did knowing the cycle length *in advance* (via Fermat/Euler) change the large-power problems compared with hunting for the cycle by hand in Chapter 9? What did the theorem buy you?

NUDGES

Launch. (1) $\varphi(12) = 12\cdot\frac{1}{2}\cdot\frac{2}{3} = 4$; $\varphi(13) = 12$; $\varphi(36) = 36\cdot\frac{1}{2}\cdot\frac{2}{3} = 12$. (2) $3^6 \equiv 1 \pmod{7}$ by Fermat. (3) Powers of 2 mod 7: $2, 4, 1$ — order 3. **Core.** (4) $3^6 \equiv 1$, $100 = 6\cdot 16 + 4$, so $3^{100} \equiv 3^4 = 81 \equiv 4 \pmod{7}$. (5) Powers of 3: $3, 2, 6, 4, 5, 1$ — order 6, which divides 6. (6) $\varphi(100) = 40$, $2024 \equiv 24 \pmod{40}$, compute $3^{24} \equiv 81 \pmod{100}$. **Stretch.** (7) $\binom{p}{k} = \frac{p!}{k!(p-k)!}$ has a factor p in the numerator uncancelled for $0 < k < p$; expanding $(\underbrace{1+\cdots+1}_{a})^p$ mod p leaves $a^p \equiv a$. (8) On each prime factor, Fermat gives $a^p \equiv a$; since $561-1$ is a multiple of each $(p-1)$, conclude via CRT. **The Flash.** (9) 49 (as in the opener). (10) $6! \equiv -1 \equiv 6 \pmod{7}$; $10! \equiv -1 \equiv 10 \pmod{11}$. **Proof.** (11) Write $m = \prod p^e$. By the Chinese Remainder Theorem a residue mod m corresponds to a tuple of residues mod each p^e, and it is a unit iff it is a unit modulo each prime power — so the coprimality conditions at distinct primes are independent and φ is multiplicative. For a single prime power the only non-units are the multiples of p, giving $\varphi(p^e) = p^e - p^{e-1} = p^e(1-1/p)$. Multiplying over all primes yields $\varphi(m) = m\prod_{p|m}(1-1/p)$. (12) In $(p-1)! = \prod_{a=1}^{p-1} a$, pair each a with its unique inverse $a^{-1} \pmod{p}$. A term is self-paired exactly when $a^2 \equiv 1$, i.e. $a \equiv \pm 1$; every other residue pairs off into a product $\equiv 1$. What survives is $1\cdot(p-1) \equiv -1 \pmod{p}$, which is Wilson's theorem.

CHAPTER ELEVEN

11

The Honeybee's Craft

A bee building a honeycomb is solving an optimization problem it never studied. The secret it has found — nature's secret — is that the best shape is almost always the most balanced one.

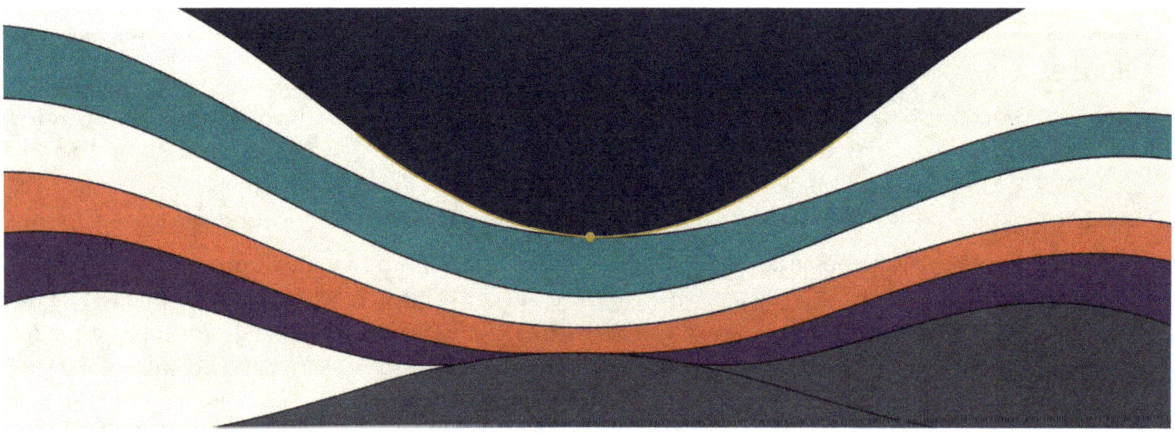

Inequalities: balance is best.

A honeybee answers the next question every day of its life.

WONDER THIS

A bee must store honey in wax, and wax is expensive — it costs the hive enormous energy to make. So the cells should hold the most honey for the least wax wall.

What shape should the cells be? Circles waste the gaps between them. Triangles and squares tile perfectly — but is one of them cheaper? And why do real honeycombs use *hexagons*?

Hold that question; we'll answer it by the end of the chapter. It turns out the bee is wrestling with the deepest theme in all of inequalities: *how to get the most from the least*. And the recurring answer (in the hive, in the soap bubble, in the farmer's field) is that nature optimizes by *balancing*. To warm up, take the schoolbook version everyone half-remembers: given 20 meters of fence, what rectangle encloses the most area? Try a few and a principle surfaces. It is one far bigger than fences. Nature does not search for the honeycomb's ideal shape; it settles into the balanced one, and the balanced one is best. Stop optimizing by brute force — start recognizing where balance must fall.

CURIOUS CARL • FIRST GUESS

Carl: "The smallest value of $x + \frac{1}{x}$? I could sample — $x = 2$, then 3, then 0.7 — and grab the lowest. Though... how would I ever be *sure* I'd hit the true floor?"

The catch: you could plug in numbers all afternoon, Carl, and never be *sure* you found the floor — but AM-GM hands you the minimum in a single line, and tells you it lives at the balance point $x = 1$.

Try a few. A 9 × 1 rectangle has area 9. A 7 × 3 has area 21. A 6 × 4 has area 24. A 5 × 5 square has area 25. The square wins. And if you test more, nothing beats it. There's a principle hiding here, one of the most useful inequalities in mathematics.

The two sides of a rectangle with perimeter 20 satisfy $a + b = 10$: their *sum* is fixed. We want to maximize their *product ab*, the area. So the real question is: *among all pairs of positive numbers with a fixed sum, which pair has the largest product?* The answer is always the pair that's equal. This is the famous **AM-GM inequality**, and for two numbers it says the arithmetic mean is never smaller than the geometric mean:

$$\frac{a+b}{2} \geq \sqrt{ab}, \qquad \text{with equality exactly when } a = b.$$

Since $\frac{a+b}{2} = 5$ is fixed, $\sqrt{ab} \leq 5$, so $ab \leq 25$. And the maximum is reached only when $a = b = 5$. **Among all rectangles of a given perimeter, the square encloses the most area.** The inequality didn't just find the answer. It told us *exactly when* equality holds, which is where the maximum lives.

THE TURN

from grind to insight

The textbook route is calculus: write the area as $a(10-a)$, differentiate, set the derivative to zero, solve for the critical point, then check it's a maximum — or, worse, just plug in dozens of values and hope the biggest one is really the biggest. It works, but it's machinery, and it never tells you *why* the answer is what it is. **The one-line escape** is to apply AM-GM once: fixed sum forces $\sqrt{ab} \leq \frac{a+b}{2}$, so the product is capped the instant you write the inequality down. It wins because the equality clause $a = b$ is baked in for free — the bound doesn't just find the maximum, it hands you the exact configuration that achieves it, no derivatives, no cases.

THE MOVE

The AM-GM Inequality. For non-negative numbers $x_1, \dots, x_n$,

$$\frac{x_1 + x_2 + \cdots + x_n}{n} \geq \sqrt[n]{x_1 x_2 \cdots x_n},$$

the arithmetic mean is at least the geometric mean, *with equality if and only if all the numbers are equal.* The "equality condition" is not a footnote — it is where optima live. To maximize a product with fixed sum (or minimize a sum with fixed product), push everything toward equal.

That equality condition is the secret heart of inequality problems, and, it turns out, the secret of the honeycomb. An inequality $A \geq B$ tells you B can't exceed A; the equality case tells you *when* B reaches that ceiling, almost always the answer to an optimization problem. The bee's instinct to make every cell identical is the equality condition, made of wax. Learn to read both halves: the bound, and the moment it's tight.

Smoothing: nudge toward balance. Here is the engine underneath "equality wins," and it needs no formula. Take any two of the positive quantities that are *unequal* and replace both by their average: their *sum* is unchanged, yet their *product* strictly *increases* (because $ab < (\frac{a+b}{2})^2$ whenever $a \neq b$). So *any* unbalanced configuration can be improved: keep smoothing unequal pairs and the product climbs until everything is equal, which must therefore be the maximum. This is why the optimum lives at the balanced point, and it often reveals the answer before you compute a thing. *Equality is the treasure map: don't search the optimum — read the equality case.*

AM-GM as a tool: bounding by balancing

The power of AM-GM is that it converts "optimize this expression" into "make these pieces equal." The craft is choosing *which* pieces to apply it to.

WATCH THIS **#1 The smallest a sum can be**

Claim. For every positive number x, the expression $x + \dfrac{1}{x}$ is at least 2, and equals 2 only when $x = 1$.

Apply AM-GM to the two terms. The two pieces are x and $\frac{1}{x}$. Their product is beautifully simple: $x \cdot \frac{1}{x} = 1$. AM-GM on these two says

$$\frac{x + \frac{1}{x}}{2} \geq \sqrt{x \cdot \tfrac{1}{x}} = \sqrt{1} = 1.$$

Multiply by 2: $x + \frac{1}{x} \geq 2$. Equality needs the two pieces equal, $x = \frac{1}{x}$, i.e. $x = 1$.

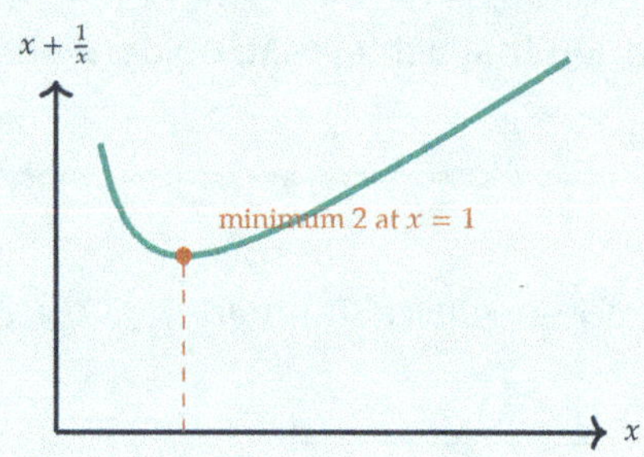

■

Trigger An expression to minimize that's a *sum* of terms, or to maximize that's a *product* — especially when the terms have a constant product or constant sum.

Do first Apply AM-GM to well-chosen terms whose product (or sum) is constant, so the bound becomes a fixed number; equality tells you where the optimum is.

Trapdoor Applying AM-GM to terms whose product *isn't* constant — then the bound still holds but isn't useful, because the right-hand side varies. Choose terms so the geometric mean is a constant.

Upgrade Split or weight terms cleverly: to minimize $x + \frac{4}{x^2}$, break x into $\frac{x}{2} + \frac{x}{2}$ so three terms have constant product.

The art, as the trapdoor below stresses, is arranging for the geometric mean to be a *constant* — that's what turns the inequality into a usable numerical bound. When you can split or group an expression so AM-GM's right side becomes a fixed number, you've found the floor (or ceiling), and the equality case hands you the optimizer for free.

△ TRAPDOOR

AM-GM only gives a *useful* bound when the geometric mean is constant. Applying it to terms whose product still depends on the variable produces a true but useless inequality. Before applying AM-GM, check: *is the product of these terms a constant?* If not, regroup, split, or weight the terms until it is. And always state the equality condition — without it you've proven a bound but not found the optimum.

CURIOUS CARL • TRAPDOOR!

Carl fires AM-GM at $\frac{a+b}{2} \geq \sqrt{ab}$ with $a = -4$, $b = -9$, declares the minimum found, and moves on. "Done — and I didn't even check whether the pieces were equal!"

The honeycomb cell tilts a warning wall: *"AM-GM only speaks for* nonnegative *numbers — feed me a negative and I'll hand you nonsense. And a bound isn't a minimum until the terms can actually be made equal."* Nice instinct, Carl — check the signs and the equality case first, and only then trust the number.

Cauchy–Schwarz: the other workhorse

The second great inequality handles *sums of products*, and it appears everywhere from geometry to probability. It looks technical but has a clean geometric soul.

WATCH THIS — #2 The Cauchy–Schwarz inequality

Claim. For any real numbers $a_1, \ldots, a_n$ and $b_1, \ldots, b_n$,

$$(a_1b_1 + a_2b_2 + \cdots + a_nb_n)^2 \leq (a_1^2 + \cdots + a_n^2)(b_1^2 + \cdots + b_n^2),$$

with equality when the two lists are proportional.

The geometric soul. Think of $\vec{a} = (a_1, \dots, a_n)$ and $\vec{b} = (b_1, \dots, b_n)$ as vectors. The left side is $(\vec{a} \cdot \vec{b})^2$ and the right is $|\vec{a}|^2|\vec{b}|^2$. Since the dot product satisfies $\vec{a} \cdot \vec{b} = |\vec{a}||\vec{b}| \cos\theta$, squaring gives

$$(\vec{a} \cdot \vec{b})^2 = |\vec{a}|^2|\vec{b}|^2 \cos^2\theta \le |\vec{a}|^2|\vec{b}|^2,$$

because $\cos^2\theta \le 1$. Equality needs $\cos^2\theta = 1$ — the vectors are *parallel*: one is a scalar multiple of the other (pointing the same way *or* exactly opposite). ■

Cauchy–Schwarz is the tool to reach for whenever you see a sum of products squared, or need to relate $\sum a_i b_i$ to the sizes of the lists. A classic consequence: setting all $b_i = 1$ gives $(\sum a_i)^2 \le n \sum a_i^2$, a bridge between a sum and a sum of squares that settles countless problems. The deeper lesson is that an algebraic inequality often has a *geometric* meaning (here, that a projection is never longer than the thing projected), and seeing the geometry makes the inequality feel inevitable rather than memorized.

Cauchy–Schwarz wears a second disguise that is so useful it gets its own name. Whenever you face a *sum of fractions whose numerators are squares*, there is a one-line move that crushes it.

WATCH THIS #3 Titu's lemma: Cauchy–Schwarz in Engel form

Claim (Titu's lemma). For real $a_1, \dots, a_n$ and *positive* $b_1, \dots, b_n$,

$$\frac{a_1^2}{b_1} + \frac{a_2^2}{b_2} + \cdots + \frac{a_n^2}{b_n} \ge \frac{(a_1 + a_2 + \cdots + a_n)^2}{b_1 + b_2 + \cdots + b_n},$$

with equality exactly when $\frac{a_1}{b_1} = \frac{a_2}{b_2} = \cdots = \frac{a_n}{b_n}$.

The two-term heart. Everything follows from the $n = 2$ case, $\frac{a^2}{x} + \frac{b^2}{y} \ge \frac{(a+b)^2}{x+y}$ for $x, y > 0$. Clear denominators: this is equivalent to $(a^2y + b^2x)(x + y) \ge (a + b)^2xy$, which expands and simplifies to $(ay - bx)^2 \ge 0$ — a square, hence always true. Equality needs $ay = bx$, i.e. $\frac{a}{x} = \frac{b}{y}$. Applying this two-term step repeatedly (or Cauchy–Schwarz on the lists $\frac{a_i}{\sqrt{b_i}}$ and $\sqrt{b_i}$) gives the full n-term form.

One clean application. For positive a, b, c,

$$\frac{a^2}{b+c} + \frac{b^2}{c+a} + \frac{c^2}{a+b} \ge \frac{(a+b+c)^2}{2(a+b+c)} = \frac{a+b+c}{2},$$

since the denominators sum to $2(a + b + c)$. **A sum that looks fearsome collapses in one line.** Equality holds when $a = b = c$. ■

Titu's lemma is the form to reach for the instant you see $\sum \frac{\text{(square)}}{\text{(positive)}}$: it folds the whole sum into a single fraction, numerators added *then* squared, denominators simply added. Many "hard" Olympiad fractions are one Titu step from done.

The showpiece: why the honeycomb is hexagonal

We can now return to the bee. It is the same story we have been telling all chapter, one step richer. Every bound so far said the same thing: *push toward balance, and read the answer off the equality case*. The honeycomb is that single principle written in wax. The cells must tile the plane with no wasted gaps, and among all tilings we want the least wall per unit of stored honey, the least *perimeter* per unit *area*. This is the *isoperimetric instinct*: nature minimizes boundary. And, exactly as with AM-GM, the winner is the most symmetric shape available.

WONDER THIS

Only three regular polygons tile the plane by themselves: the triangle, the square, and the hexagon.

Which one walls off a given area most cheaply? And how close does the winner come to the perfect shape, the circle?

WATCH THIS — **#4 The most economical tile**

The ideal, ignoring tiling: the circle. Among *all* shapes enclosing a fixed area A, the circle has the least perimeter — the *isoperimetric inequality*, $P^2 \geq 4\pi A$, equality only for the circle. A single soap bubble obeys it: surface tension pulls it to a sphere, the least-area way to wrap a fixed volume. **Left alone, nature rounds things off.** But circles leave gaps when packed, so the bee cannot use them.

The contest among tilers. Scale each cell to enclose unit area and compare perimeters. For a regular n-gon of unit area, a short computation gives perimeter $P_n = \sqrt{4n\tan(\pi/n)}$:

triangle $P \approx 4.559$ > **square** $P = 4.000$ > **hexagon** $P \approx 3.722$ > **circle** $P \approx 3.545$

The hexagon's perimeter, ≈ 3.722, beats the square's 4 and the triangle's ≈ 4.559, and sits closest to the circle's unreachable ideal $2\sqrt{\pi} \approx 3.545$. **Of the three shapes that tile, the hexagon walls off area most cheaply.** (That the hexagon beats *every* tiling, not merely the three regular ones, is the far deeper *honeycomb theorem*, proved only in 1999.) ■

perimeter per unit area (equal-area cells)

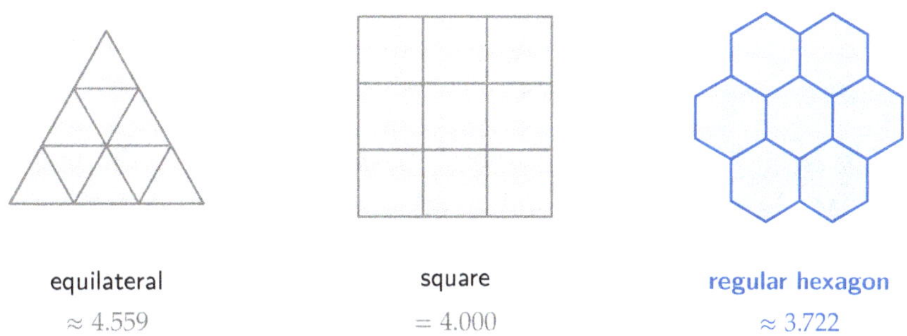

least perimeter per area → the bee's choice

Among the regular tilings, the hexagon encloses each unit of area with the least perimeter — the honeybee's choice.

So the bee's hexagon is the best compromise between two competing demands: be round (to minimize wall, like a bubble), but also tile without gaps (to waste no space). The hexagon is as close to a circle as a gap-free tile can get. That this is *truly* optimal among *all* ways to tile the plane (not just the regular ones) is the *Honeycomb Conjecture*, stated by ancients and proved only in 1999 by Thomas Hales, and only with pages of modern mathematics. Yet the bee had the answer long before any of it: no calculus, no proof, not even a word for "perimeter," just tens of millions of years of evolution quietly shaving the wax it takes to store the honey. A gram of insect settled an optimization problem humanity could not confirm until the final year of the twentieth century. Let that sit for a moment before you read on.

TRY TO BREAK IT

AM-GM insists $\frac{a+b}{2} \geq \sqrt{ab}$ — but it quietly assumes a and b are *nonnegative.* Feed it a negative value and watch the inequality lie: what is the smallest, simplest pair (a, b) you can find for which $\frac{a+b}{2} \geq \sqrt{ab}$ fails outright? (*Try one negative number and one positive — and notice what* $\sqrt{ab}$ *is even allowed to* mean.) When you find it, you'll feel in your bones why "nonnegative" is not a footnote but the load-bearing wall.

SEEN BEFORE?

The "round is cheapest" isoperimetric instinct is the geometric face of every optimization in this chapter: a quantity is extremal exactly at the balanced configuration. **Chapter 14** exploits the same idea — shortest paths and reflections find an optimum by a clever construction rather than calculus — and Cauchy–Schwarz's vector view previews the coordinate methods of **Chapter 13**.

The ladder of means

AM-GM compares two particular averages. But there's a whole family of "means," and they line up in a fixed order that never varies, the mechanism behind nearly every elementary inequality.

For positive a and b, there are several ways to average: the ordinary (arithmetic), the root-of-the-product (geometric), the reciprocal-of-the-average-of-reciprocals (harmonic), and the root-mean-square (quadratic). Do they have a fixed pecking order?

WATCH THIS **#5 Quadratic ≥ Arithmetic ≥ Geometric ≥ Harmonic**

Claim. $\sqrt{\dfrac{a^2+b^2}{2}} \geq \dfrac{a+b}{2} \geq \sqrt{ab} \geq \dfrac{2}{\frac{1}{a}+\frac{1}{b}}.$

The middle link (AM ≥ GM) we already have.

The left link (QM ≥ AM) follows because $a^2+b^2 \geq \frac{(a+b)^2}{2}$, which rearranges to $(a-b)^2 \geq 0$ — always true.

The right link (GM ≥ HM) is AM-GM in disguise: the harmonic mean is $\frac{2ab}{a+b}$, and $\sqrt{ab} \geq \frac{2ab}{a+b}$ rearranges to $a+b \geq 2\sqrt{ab}$ — AM-GM again.

$$\boxed{\text{QM}} \geq \boxed{\text{AM}} \geq \boxed{\text{GM}} \geq \boxed{\text{HM}}$$

all equal $\iff a = b$

The unifying equality case. Every one of these inequalities becomes an equality at exactly the same moment: when $a = b$. The whole ladder collapses to a single point precisely when the two numbers coincide. For $a = 3, b = 7$: QM $\approx$ 5.39, AM = 5, GM $\approx$ 4.58, HM = 4.2 — strictly descending, as promised. ∎

The means aren't a random list. They're a totally ordered hierarchy, each rung provable from $(a-b)^2 \geq 0$ or AM-GM, all sharing the single equality condition $a = b$. That recurring "$(\text{something})^2 \geq 0$" is the bedrock under almost every elementary inequality: squares are never negative, and an astonishing amount of mathematics is just that fact, cleverly dressed. When stuck on an inequality, a good first instinct is to ask what square you can force to be non-negative.

Why the ladder leans: the shape of a curve

There's a single idea that explains *why* unbalancing the inputs always pushes these averages apart in the same direction — and it's about the *shape* of a function. A function is **convex** if its graph bends upward, like a valley ($\smile$): a stretched rope, the curve $y = x^2$, the bowl $y = 1/x$ for $x > 0$. The defining feature is that the chord between any two points on the graph sits *above* the curve.

THE MOVE

Jensen's inequality (the conceptual umbrella). For a convex (upward-bending) function f, the *average of the outputs* is at least the *output of the average*:

$$\frac{f(x_1) + \cdots + f(x_n)}{n} \geq f\left(\frac{x_1 + \cdots + x_n}{n}\right),$$

with equality when all the x_i are equal — and, if f is *strictly* convex, *only* then. (A merely convex function can hit equality more freely: an affine $f(x) = mx + b$ gives equality for *every* choice of inputs, since averaging commutes with a straight line.) For a concave, downward-bending function $\frown$, the inequality flips. Spreading the inputs out can only *raise* the average height on a valley-shaped curve.

This is the hidden engine behind the whole mean ladder. QM $\geq$ AM is exactly Jensen applied to the convex function $f(x) = x^2$: the mean of the squares beats the square of the mean. AM $\geq$ GM is Jensen applied to the concave log (taking logs turns the geometric mean into an average). The reason every average in the ladder slides apart the instant you unbalance the inputs is that each is secretly f(average) versus average of f for some curved f. And a curve, by definition, separates those two. **Balance wins because curves bend the same way everywhere.** *The mean ladder isn't four facts; it's one fact about valleys, viewed through four windows.*

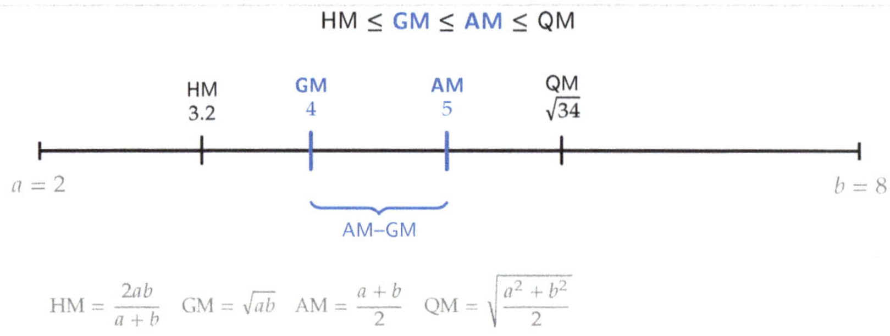

The four means of $a = 2$ and $b = 8$, drawn to scale: $\mathrm{HM} \le \mathrm{GM} \le \mathrm{AM} \le \mathrm{QM}$.

Order matters: the rearrangement inequality

Here is a different flavor of balancing. Suppose you must pair up two lists of numbers (say wages $\$1, \$2, \$3$ with hours $4, 5, 6$) and your total cost is the sum of the products. *How you match them* changes the total. Pairing big with big inflates the sum; pairing big with small deflates it. This simple, almost obvious fact is a surprisingly sharp tool.

WATCH THIS — **#6 The rearrangement inequality**

Claim. Let $a_1 \le a_2 \le \cdots \le a_n$ and $b_1 \le b_2 \le \cdots \le b_n$ be two sorted lists, and let σ be any permutation. Then

$$\underbrace{\sum_i a_i b_{n+1-i}}_{\text{oppositely sorted}} \;\le\; \sum_i a_i b_{\sigma(i)} \;\le\; \underbrace{\sum_i a_i b_i}_{\text{same sorted}} .$$

The sum of paired products is *largest* when both lists are sorted the same way, and *smallest* when sorted oppositely.

The swap argument. Suppose a pairing is *not* sorted the same way: somewhere a larger a is matched to a smaller b while a smaller a is matched to a larger b. Say $a_i < a_j$ but they are paired with $b_p > b_q$. Swapping the two partners changes the sum by

$$(a_i b_q + a_j b_p) - (a_i b_p + a_j b_q) = (a_j - a_i)(b_p - b_q) \ge 0,$$

a product of two non-negative factors. So every swap toward "same-sorted" can only *raise* the sum — and the maximum must be the fully same-sorted pairing. Running the argument with the inequality reversed gives the oppositely-sorted minimum. **Like with like is biggest; like with unlike is smallest.**

One clean application. Take the list $a = (a, b, c)$ paired against itself. The same-sorted pairing gives $a^2 + b^2 + c^2$; *any* other pairing, such as $ab + bc + ca$, can be no larger:

$$a^2 + b^2 + c^2 \;\ge\; ab + bc + ca.$$

The same idea proves $\frac{a}{b} + \frac{b}{c} + \frac{c}{a} \ge 3$ for positive a, b, c: pairing each of a, b, c with its *own* reciprocal gives $\frac{a}{a} + \frac{b}{b} + \frac{c}{c} = 3$, which is the oppositely-sorted (hence *minimum*) pairing, so any other pairing such as $\frac{a}{b} + \frac{b}{c} + \frac{c}{a}$ is at least 3. ■

Trigger A sum of *paired products* (or paired ratios) where you control how two lists are matched, and you must bound the sum — or a symmetric inequality comparing $\sum a_i b_i$ against a "scrambled" version like $\sum a_i b_{\sigma(i)}$.

Do first Sort both lists. Decide whether you want the maximum (sort them the same way) or the minimum (sort them oppositely), then compare the given pairing to that extreme.

Trapdoor Forgetting the lists must be *sorted* before reading off "same-sorted is largest" — the inequality is about sorted order, not the order the numbers happen to appear in. Sort first.

Upgrade Apply it to a list paired *against itself* (or against its own reciprocals) to manufacture classics like $a^2 + b^2 + c^2 \geq ab + bc + ca$ and $\frac{a}{b} + \frac{b}{c} + \frac{c}{a} \geq 3$ in one stroke.

The rearrangement inequality is the combinatorial cousin of AM-GM: where AM-GM says "equal inputs win," rearrangement says "aligned pairings win." Both are really the statement that a single well-chosen square (here $(a_j - a_i)(b_p - b_q) \geq 0$) is non-negative. *Even the order you hold things in is an optimization.*

Payoff: the most balanced box

Let's close by lifting the garden one dimension higher — and watch balance win again.

WONDER THIS

The three dimensions of a box must add to 12.

What box has the largest volume? And do you already suspect the answer?

If the rectangle and the honeycomb taught you anything, you can guess: the most balanced box (the cube) should win. AM-GM proves it instantly. With $x + y + z = 12$ fixed, the three-variable AM-GM says

$$\frac{x+y+z}{3} \geq \sqrt[3]{xyz} \quad \Longrightarrow \quad 4 \geq \sqrt[3]{xyz} \quad \Longrightarrow \quad xyz \leq 64,$$

with equality exactly when $x = y = z = 4$. **Among all boxes with a fixed total edge-length, the cube holds the most volume** — the same balancing principle as the square and the bee's hexagon, lifted one dimension higher. *Nature, it seems, has a soft spot for things that are equal to themselves — and so does the arithmetic mean.* The single idea "fixed sum, maximize product, push toward equal" unlocks the garden, the box, the bubble, and the honeycomb alike.

What to try when you're stuck

When a problem asks you to prove one quantity is at least (or at most) another, or to find a maximum or minimum:

1. **Reach for AM-GM** when optimizing a product with fixed sum (or vice versa): the optimum is where everything's equal.
2. **Make the geometric mean constant.** Split, group, or weight terms so AM-GM's right side is a fixed number — otherwise the bound is useless.

3. **Reach for Cauchy–Schwarz** when you see a sum of products, or need to relate $\sum a_ib_i$ to $\sum a_i^2$ and $\sum b_i^2$.

4. **Force a square.** Many inequalities reduce to (something)$^2 \geq 0$. Ask what square makes your claim obvious.

5. **Trust the isoperimetric instinct.** When a shape or configuration is free to vary, the optimum is usually the most symmetric one — round, equal, balanced.

6. **Always find the equality case.** It's almost always where the optimum lives — and stating it completes the proof.

THE MOVE

One-page recap. An inequality has two halves: the bound, and *when it's tight*. Toolbox: **AM-GM** (mean $\geq$ geometric mean, equal when all equal — maximize products with fixed sum), **make the GM constant** (split/group terms), **Cauchy–Schwarz** ($(\sum a_ib_i)^2 \leq \sum a_i^2 \sum b_i^2$, a projection bound), **force a square** ($(\cdot)^2 \geq 0$), and the **isoperimetric instinct** (round/balanced shapes minimize boundary). The mean ladder QM $\geq$ AM $\geq$ GM $\geq$ HM collapses to equality exactly when all values agree. When optimizing, push toward balance.

THE NAME FOR IT

The classical inequalities

The bound $\frac{a_1+\cdots+a_n}{n} \geq \sqrt[n]{a_1\cdots a_n}$ is the *AM–GM inequality*; the projection bound $(\sum a_ib_i)^2 \leq (\sum a_i^2)(\sum b_i^2)$ is the *Cauchy–Schwarz inequality*; their full hierarchy is the *QM–AM–GM–HM mean chain*. "Least perimeter for fixed area" is the *isoperimetric inequality*.

WHERE IT LEADS AM–GM and Cauchy–Schwarz are the elementary tip of *convex analysis* and *inner-product (Hilbert) space* geometry: Cauchy–Schwarz is the defining inequality of every inner product, and *Jensen's inequality* generalizes AM–GM to all convex functions. The honeycomb's optimality is the *isoperimetric problem*, whose discrete plane version (the *Honeycomb Conjecture*) was proved by Hales in 1999 — the bedrock of optimization, the calculus of variations, and mathematical economics.

CURIOUS CARL • EUREKA

Carl pokes one more value, gives up, and watches the scale level itself. "The winner was sitting at the balance point being smug about it. I brought calculus to a fight I could've won by holding still."

SAME BEAST, NEW COSTUME

The same "push toward balance, and one bound finds the optimum" beast, in disguise across the book:

AM–GM → Cauchy–Schwarz → Titu's lemma → Rearrangement → The honeycomb.

Each squeezes an expression against a fixed bound and reads off the answer at the balanced, aligned, or equal configuration — the equality case *is* the optimum.

The practice ladder

THE PRACTICE LADDER *climb at your own pace*

LAUNCH *Warm the engine. Nudges below.*

1. Among rectangles with perimeter 36, which has the largest area, and what is it?

2. Prove $x + \frac{1}{x} \geq 2$ for $x > 0$ and state when equality holds.

CORE BUILDER *Master the central moves.*

1. Find the minimum of $x + \frac{9}{x}$ for $x > 0$, with the value of x that achieves it.

2. Prove that for positive a, b, c with $a + b + c = 1$, we have $\frac{1}{a} + \frac{1}{b} + \frac{1}{c} \geq 9$.

3. Use Cauchy–Schwarz to prove $(a + b + c)^2 \leq 3(a^2 + b^2 + c^2)$ and state the equality case.

STRETCH *Transfer to unfamiliar ground.*

1. Find the maximum of xyz subject to $x + y + z = 12$ with $x, y, z > 0$, and prove it with AM-GM.

2. Prove the rearrangement-flavored fact $a^2 + b^2 + c^2 \geq ab + bc + ca$ for all reals by forcing a sum of squares.

3. (*Titu's lemma.*) For positive a, b, c, prove $\dfrac{a^2}{b+c} + \dfrac{b^2}{c+a} + \dfrac{c^2}{a+b} \geq \dfrac{a+b+c}{2}$, and state when equality holds.

THE FLASH *No clock — each cracks open in one flash of insight, once you find the angle. Nudges at the back.*

1. What is the minimum of $\frac{a}{b} + \frac{b}{a}$ for positive a, b?

2. For $a = 4, b = 9$, compute QM, AM, GM, HM and confirm the ordering.

PROOF LAB *Write it clean, in full sentences.*

1. Prove the two-variable AM-GM $\frac{a+b}{2} \geq \sqrt{ab}$ by forcing a square, and prove the full mean ladder QM $\geq$ AM $\geq$ GM $\geq$ HM with the shared equality condition.

2. Prove the n-variable AM-GM by any method you like (induction, or the "smoothing" argument that replaces the largest and smallest by their average), and identify exactly when equality holds.

3. (*Rearrangement.*) State the rearrangement inequality for sorted lists, prove it by the swap (exchange) argument, and use it to prove $\frac{a}{b} + \frac{b}{c} + \frac{c}{a} \geq 3$ for positive a, b, c.

REFLECTION *No nudge. Write a few sentences.*

1. The honeybee, the soap bubble, and the cube all "solve" an optimization by being balanced. Pick one optimization in this chapter and describe how the equality case pointed to the answer *before* the algebra finished.

NUDGES

Launch. (1) The 9×9 square, area 81. (2) AM-GM on $x, \frac{1}{x}$ whose product is 1; equality at $x = 1$. **Core.** (3) AM-GM gives $x + \frac{9}{x} \geq 2\sqrt{9} = 6$, at $x = 3$. (4) By AM-HM (or Cauchy), $(\frac{1}{a} + \frac{1}{b} + \frac{1}{c})(a + b + c) \geq 9$; since $a + b + c = 1$, done. (5) Cauchy with $b_i = 1$; equality when $a = b = c$. **Stretch.** (6) AM-GM gives $\sqrt[3]{xyz} \leq 4$, so $xyz \leq 64$ at $x = y = z = 4$. (7) $a^2 + b^2 + c^2 - ab - bc - ca = \frac{1}{2}((a - b)^2 + (b - c)^2 + (c - a)^2) \geq 0$. (8) Titu's lemma: the left side $\geq \frac{(a+b+c)^2}{2(a+b+c)} = \frac{a+b+c}{2}$; equality at $a = b = c$. **The Flash.** (9) Minimum 2, at $a = b$. (10) QM ≈ 6.96, AM $= 6.5$, GM $= 6$, HM ≈ 5.54; strictly descending. **Proof.** (11) Since $\frac{a+b}{2} - \sqrt{ab} = \frac{1}{2}(\sqrt{a} - \sqrt{b})^2 \geq 0$, two-variable AM-GM holds with equality iff $a = b$. For the ladder, QM $\geq$ AM is $2(a^2 + b^2) \geq (a + b)^2$, i.e. $(a - b)^2 \geq 0$; AM $\geq$ GM is the square just shown; and GM $\geq$ HM rewrites as $\sqrt{ab}\,(a + b) \geq 2ab$, which is AM-GM once more. Every link is a squared difference ≥ 0, so the four means coincide precisely when $a = b$. (12) Fix the sum and watch the product: if two entries differ, replacing both by their average leaves the sum unchanged but strictly raises the product, since $\left(\frac{x+y}{2}\right)^2 > xy$. Iterating drives every entry to the common value $\frac{1}{n}\sum a_i$, where the product is maximal and equals $(\frac{1}{n}\sum a_i)^n$; hence $\frac{1}{n}\sum a_i \geq \sqrt[n]{\prod a_i}$, with equality iff all a_i are equal. (13) For sorted lists, any pairing that is not similarly sorted has indices with $a_i < a_j$ matched against $b_p > b_q$; swapping their partners changes the dot product by $(a_j - a_i)(b_p - b_q) \geq 0$, so the similarly-sorted pairing is largest and the oppositely-sorted one smallest. Pairing (a, b, c) against $(\frac{1}{a}, \frac{1}{b}, \frac{1}{c})$, the oppositely-sorted (minimum) arrangement matches each term with its own reciprocal, giving $\frac{a}{a} + \frac{b}{b} + \frac{c}{c} = 3$; therefore $\frac{a}{b} + \frac{b}{c} + \frac{c}{a} \geq 3$.

CHAPTER TWELVE

12

The Back Door to the Imaginary

A formula for the cubic was the first thing algebra could do that geometry could not — and its strangest case forced mathematicians to compute with $\sqrt{-1}$, a number they had spent a century swearing did not exist. This is the back door through which the imaginary numbers walked in.

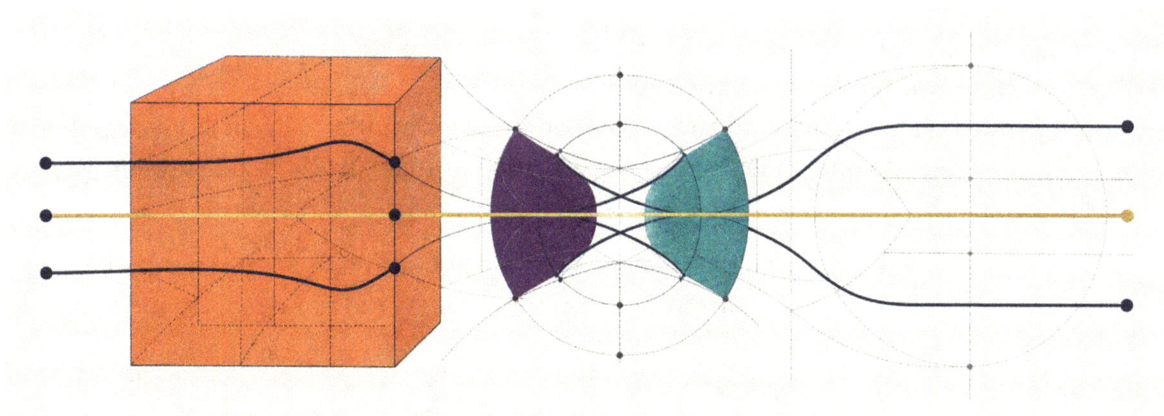

To crack a cubic: guess the shape, and split one unknown into two.

WONDER THIS

Simplify

$$\sqrt[3]{10 + 6\sqrt{3}} + \sqrt[3]{10 - 6\sqrt{3}}.$$

The answer is a small positive integer. **Which one — and why must a monster of nested radicals collapse to something so tame?** By the chapter's end you will find it in three lines, and you will know why an expression like this *had* to be invented, who nearly died keeping it secret, and how it forced mathematicians to accept numbers they spent a century calling "impossible."

CURIOUS CARL • FIRST GUESS

Carl: "I'll punch the cube roots into a calculator — $\sqrt[3]{10+6\sqrt{3}} \approx 2.73$, $\sqrt[3]{10-6\sqrt{3}} \approx -0.73$, sum about 2. So... the answer is just 2? That feels too clean."

The catch: a decimal that *looks* like 2 isn't a proof, Carl — and worse, the calculator hides the real prize. The whole reason this expression exists is that it is secretly an integer, and *seeing why* unlocks a five-hundred-year-old story about the birth of algebra. Grind the decimals and you learn nothing; find the structure and you learn everything.

Behind this innocent puzzle stands the most dramatic episode in the history of algebra: a public duel, a broken oath, and a scandal over "impossible" numbers that would not resolve for three hundred years. Let us take it in order. This is the boldest escape in the book: **don't win the fight you were handed — leave the world the problem lives in, solve it in a friendlier one, and carry the answer home.**

A duel in Venice

Mathematics has produced few scenes as cinematic as the one that unfolded in Venice on February 22, 1535.

Two men faced each other in a public contest. Each had deposited thirty problems with a notary; each had forty days to solve the other's thirty. The stakes were real: in Renaissance Italy a mathematician's income depended on his reputation, and reputation was won and lost in exactly these public duels. Lose badly enough and your students left, your patrons vanished, and your lectureship went to the winner.

On one side stood Antonio Fior, a mediocre mathematician holding one magnificent secret. His teacher, Scipione del Ferro of Bologna, had done what the entire ancient world could not: around 1515 he found a method for solving cubic equations of the form

$$x^3 + px = q \qquad \text{("cube plus things equals a number").}$$

Del Ferro never published it. Following the custom of the age, he guarded it as a professional weapon and revealed it on his deathbed in 1526 to a few confidants, Fior among them. Nine years later Fior decided to cash in his inheritance. All thirty of his challenge problems were cubics of that one form: a single bullet, fired thirty times.

On the other side stood Niccolò Fontana of Brescia, known to history as **Tartaglia**: "the stammerer," and the nickname was a scar. As a boy he was among the townspeople cut down when French troops sacked Brescia in 1512; a soldier's saber split his jaw and palate, and though he survived, he never spoke cleanly again. Too poor for a proper schooling (by legend he practiced his letters on tombstones for want of paper), he made himself into a mathematician by sheer will. He had heard only a *rumor* that the cubic could be solved, but rumor is a powerful hint: knowing that a problem *can* be cracked is half the battle. In the final days before the deadline (by his own account, on the night of February 12–13, 1535) he found the method himself. On the day of reckoning he tore through all thirty of Fior's problems in roughly two hours; Fior, facing Tartaglia's more varied list, solved almost none. It was a rout.

The oath and the betrayal

News of the duel reached Girolamo Cardano in Milan: physician, gambler, astrologer, and one of the most brilliant and strange minds of the century. Cardano begged Tartaglia for the method. In 1539, after long resistance, Tartaglia relented and disclosed it, encoded in twenty-five lines of verse, under a sworn oath of secrecy: "I swear to you by the Sacred Gospels... never to publish your discoveries, if you tell them to me."

Cardano kept the oath for four years. But the poem gave only the recipe, not the reasoning; Cardano and his prodigious student Lodovico Ferrari reconstructed the proofs, extended the method to *all* cubics, and Ferrari — still a teenager — solved the quartic. Then, in 1543, the two traveled to Bologna and examined del Ferro's surviving papers. There it was, in the dead man's own hand, dated years before Tartaglia's discovery. Cardano reasoned that his oath bound him to Tartaglia's secret — not del Ferro's.

In 1545 he published *Ars Magna* ("The Great Art"), crediting both del Ferro and Tartaglia by name and laying out the complete theory of cubic and quartic equations. Tartaglia was incandescent. The feud raged in pamphlets for years and culminated in a public debate in Milan in 1548 against Ferrari, which Tartaglia lost. He died poor. Cardano's formula, which he did not invent alone and Tartaglia did not invent first, carries Cardano's name to this day.

Why does this melodrama matter to us? Because *Ars Magna* of 1545 is the most defensible birthday of modern algebra. Before it, algebra meant inherited recipes; after it, algebra became a research frontier: a subject where new methods could be *discovered*, where formulas manipulated symbols beyond geometric intuition, and where (as we are about to see) the manipulations themselves forced entirely new numbers into existence.

Three thousand years of waiting

To feel the weight of 1545, look at how long the world had been stuck.

Babylon, c. 1800–1600 BC. Clay tablets show scribes solving problems we would write as $x^2 + bx = c$, using a procedure exactly equivalent to completing the square. They had no symbols (everything was prose and worked examples), but the method was correct, systematic, and taught in schools. The quadratic equation is *older than the alphabet*.

Baghdad, c. 820 AD. Al-Khwarizmi's treatise *al-Kitab al-mukhtasar fi hisab al-jabr wa'l-muqabala* organized quadratic equations into cases and solved each with geometric justification. The word *al-jabr* ("restoration") became our word **algebra**; the author's name, Latinized, became **algorithm**. Masterful, rigorous — and still quadratic.

Persia, c. 1070. Omar Khayyam (yes, the poet of the *Rubaiyat*) classified cubic equations into fourteen types and solved each *geometrically*, as the intersection of two conic sections: a circle meeting a parabola, a hyperbola meeting a circle. It was beautiful and it worked, but it produced a *point on a drawing*, not a formula. Khayyam himself wrote that an arithmetic solution of the cubic was beyond him and might be found by "those who come after us."

They came after him by four and a half centuries.

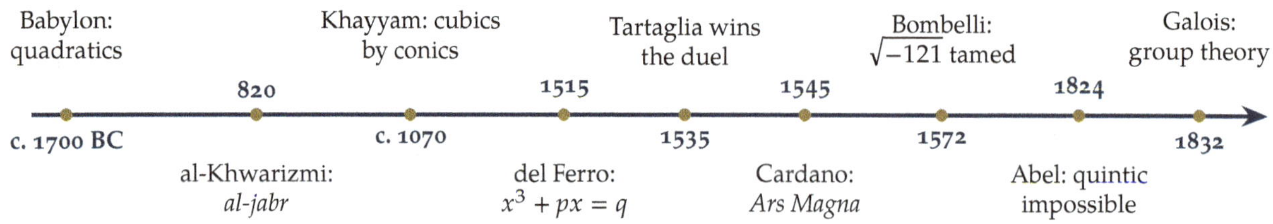

Three millennia from the quadratic to the cubic; then only 287 years from the cubic to Galois. Solving the cubic didn't just answer a question. It changed the *speed* of mathematics. Why was the cubic so much harder? One honest answer: completing the square has a perfect geometric picture (an actual square, completed by an actual corner piece), and "completing the cube" has no such picture that finishes the job. The cubic needed an idea that was *algebraic from birth*: a leap of symbolic imagination with no geometric safety net. That is precisely why we date modern algebra from its solution.

Step one: depressing the cubic

The general cubic is $x^3 + ax^2 + bx + c = 0$. The first move (known already to Cardano) is to kill the x^2 term. Substitute

$$x = t - \frac{a}{3}.$$

Why $a/3$? The roots of a cubic have centroid $-a/3$ (their sum is $-a$ by Vieta, so their average is $-a/3$). The substitution slides the graph sideways so the roots are centered at 0; the resulting cubic, having roots summing to zero, has no square term. A cubic with no square term is called **depressed**.

WATCH THIS **#1 Depressing a cubic**

Task. Depress $x^3 - 3x^2 - 3x - 1 = 0$.
Do it. Here $a = -3$, so substitute $x = t + 1$:

$$(t+1)^3 - 3(t+1)^2 - 3(t+1) - 1 = t^3 - 6t - 6.$$

So we must solve $t^3 - 6t - 6 = 0$, then recover $x = t + 1$. (Hold this example — we will finish it shortly, and its answer is a gem.)

From now on we work with the depressed form

$$\boxed{t^3 + pt + q = 0.}$$

The two-unknowns gambit

Everything before this paragraph, a Babylonian could have followed. What comes next is the birth certificate of modern algebra, and the signature creative move of the chapter.

THE TURN *from grind to insight*

The grind. You want to "complete the cube" the way you complete the square. But there is no corner piece: no rearrangement of $t^3 + pt + q$ into a perfect cube plus a constant. Geometry has run out of pictures.

The turn. Stop trying to simplify one unknown; *split* it into two. Guess the *shape* of the answer:

$$t = u + v.$$

This looks like a step backwards — we traded one unknown for two! But two unknowns carry a spare degree of freedom, and we will spend it to make the equation collapse. Expand with the identity $(u + v)^3 = u^3 + v^3 + 3uv(u + v)$:

$$(u + v)^3 + p(u + v) + q = u^3 + v^3 + (3uv + p)(u + v) + q = 0.$$

Now spend the freedom: **demand** that $3uv + p = 0$, i.e. $uv = -\frac{p}{3}$. The middle term vanishes, leaving $u^3 + v^3 = -q$. So we need two numbers u^3, v^3 with

$$u^3 + v^3 = -q, \qquad u^3v^3 = -\frac{p^3}{27}.$$

Known sum, known product — so u^3 and v^3 are the two roots of a *quadratic*, which Babylon could already solve:

$$z^2 + qz - \frac{p^3}{27} = 0.$$

The cubic has collapsed into a quadratic, at the price of a cube root at the end.

Solving that quadratic and taking cube roots gives **Cardano's formula** — a square root nested inside two cube roots:

$$\boxed{t = \sqrt[3]{-\frac{q}{2} + \sqrt{\frac{q^2}{4} + \frac{p^3}{27}}} + \sqrt[3]{-\frac{q}{2} - \sqrt{\frac{q^2}{4} + \frac{p^3}{27}}}}$$

Pause on the logic of the gambit, because it is the deepest lesson here. The same shape-guessing tames an ordinary nested *square* root: to simplify $\sqrt{a + b\sqrt{c}}$, guess it equals $\sqrt{p} + \sqrt{q}$, square, and match the rational and irrational parts to get a sum-and-product system — again a quadratic. Cardano's method and radical *denesting* are not merely similar; they are the same idea running in opposite directions, and we will drive on that two-way street for the rest of the chapter.

THE MOVE

Equation ⇌ roots. Do not read Cardano's derivation as a one-way trick for extracting roots from equations. Read the key identity $(u+v)^3 = u^3+v^3+3uv(u+v)$ as a statement about the *number* $x = u+v$: it says x satisfies the cubic $x^3 - 3uv\,x - (u^3+v^3) = 0$. So there are two directions, and fluency means travelling both:

- **Equation → roots** (Cardano): given p, q, manufacture u, v and write the root as nested radicals.
- **Roots → equation** (the reverse): given a nasty number $x = \sqrt[3]{A}+\sqrt[3]{B}$, read off the clean cubic it satisfies, find that root by easier means (rational root test!), and conclude the monster equals something simple.

A number and its equation are two descriptions of one object. *Think of the expression as a root; think of the root through its equation.* This single habit powers half of olympiad algebra — and, scaled up, it becomes Galois theory.

The formula speaks in riddles

Let us take the formula for a drive — straight into the Opening Challenge.

WATCH THIS #2 Fior's favorite type

Solve $x^3 + 6x - 20 = 0$.
Do it. Here $p = 6, q = -20$, so $\frac{q^2}{4} + \frac{p^3}{27} = 100 + 8 = 108$ and $\sqrt{108} = 6\sqrt{3}$. Cardano announces

$$x = \sqrt[3]{10+6\sqrt{3}} + \sqrt[3]{10-6\sqrt{3}}.$$

This is exactly our Opening Challenge — and here is the punchline. A glance (or the rational root test) shows $x = 2$ works: $8 + 12 - 20 = 0$, and since $x^3 + 6x$ is strictly increasing, 2 is the *only* real root. So this monster of nested radicals **must equal** 2. The formula is correct — it just speaks in riddles.

Confirm it by denesting. Guess $\sqrt[3]{10+6\sqrt{3}} = m + n\sqrt{3}$; cubing and matching parts (or testing small values) gives $(1+\sqrt{3})^3 = 10+6\sqrt{3}$. For the second radical, note that $10 - 6\sqrt{3} \approx -0.39$ is *negative*; its real cube root is the negative number $1-\sqrt{3}$, since $(1-\sqrt{3})^3 = 10-6\sqrt{3}$. Therefore

$$x = (1+\sqrt{3}) + (1-\sqrt{3}) = 2. \qquad \checkmark$$

△ TRAPDOOR

When you denest a *cube* root, mind the sign. Unlike square roots, cube roots of negative numbers are perfectly legal and perfectly negative: $\sqrt[3]{10-6\sqrt{3}} = 1-\sqrt{3} < 0$, not $\sqrt{3}-1$. Guess the wrong sign and your two radicals will add to the wrong integer.

Now we can finish the depressed example from before in style.

WATCH THIS #3 A gem hiding in nested radicals

Solve $x^3 - 3x^2 - 3x - 1 = 0$ exactly.
Do it. We depressed it to $t^3 - 6t - 6 = 0$, so $p = -6$, $q = -6$:

$$\frac{q^2}{4} + \frac{p^3}{27} = 9 - 8 = 1, \qquad t = \sqrt[3]{3+1} + \sqrt[3]{3-1} = \sqrt[3]{4} + \sqrt[3]{2}.$$

This time the cube roots *cannot* be denested — the root really is irrational. Restoring $x = t + 1$ gives $x = 1 + \sqrt[3]{2} + \sqrt[3]{4}$. And here is the gem: that is a geometric series in $\sqrt[3]{2}$, so

$$x = \frac{(\sqrt[3]{2})^3 - 1}{\sqrt[3]{2} - 1} = \frac{1}{\sqrt[3]{2} - 1}.$$

The unique real root of $x^3 - 3x^2 - 3x - 1 = 0$ is the reciprocal of $\sqrt[3]{2} - 1$. (Multiply $(1 + \sqrt[3]{2} + \sqrt[3]{4})(\sqrt[3]{2} - 1) = 2 - 1 = 1$ to see it in one line.)

SEEN BEFORE?

When does Cardano's formula give a "nice" answer? Fior's thirty duel problems all had simple answers, yet the formula dresses them in nested radicals. The pattern: *if the depressed cubic has a rational root, the radicals denest* (as in $x^3 + 6x - 20$); *if it has no rational root, no denesting exists* (as in $t^3 - 6t - 6$) and the nested radical is already the simplest possible name for that number. So whether a radical denests is a question about the roots of its hidden equation — the equation⇌roots move doing real work, not a manipulation drill.

Reading Cardano backwards

Competition writers love the reverse direction: they hand you the nested radicals and hide the cubic. Your job is to rebuild it.

THE MOVE

From a nested radical to its cubic. Given $x = \sqrt[3]{A} + \sqrt[3]{B}$, set $s = A + B$ and $P = \sqrt[3]{AB}$. Cubing via $(u+v)^3 = u^3 + v^3 + 3uv(u+v)$:

$$x^3 = s + 3Px \qquad \Longrightarrow \qquad x^3 - 3Px - s = 0.$$

If AB is a perfect cube (the problems are built so it is), this is a clean cubic with rational coefficients. Find its rational root; check it is the only real one (automatic when $P \le 0$, since then the cubic is increasing); done.

WATCH THIS **#4 Rebuild the equation, read off the answer**

Simplify $\sqrt[3]{18+5\sqrt{13}}+\sqrt[3]{18-5\sqrt{13}}$.
Do it. Here $s = 36$ and $AB = 18^2 - 25\cdot 13 = 324 - 325 = -1$, so $P = \sqrt[3]{-1} = -1$. Thus x satisfies

$$x^3 = 36 - 3x \quad\Longleftrightarrow\quad x^3 + 3x - 36 = 0.$$

The rational root test finds $x = 3$ $(27 + 9 - 36 = 0)$, and $x^3 + 3x$ is strictly increasing, so there are no other real roots. The expression equals $\boxed{3}$.

There is an even more revealing version, where the answer stays hidden on purpose.

WATCH THIS **#5 Going in a circle — on purpose**

Simplify $\sqrt[3]{n+\sqrt{n^2+8}}+\sqrt[3]{n-\sqrt{n^2+8}}$ for a positive integer n.
Do it. Now $s = 2n$ and $AB = n^2 - (n^2+8) = -8$, so $P = -2$ and

$$x^3 = 2n - 6x \quad\Longleftrightarrow\quad x^3 + 6x - 2n = 0.$$

Since $x^3 + 6x$ is strictly increasing, the expression is *the unique real root* of $x^3 + 6x = 2n$ — and in general that root is irrational. But look what we have written: Cardano's formula applied to $x^3 + 6x - 2n = 0$ (where $\frac{q^2}{4} + \frac{p^3}{27} = n^2 + 8$) gives back exactly $\sqrt[3]{n+\sqrt{n^2+8}}+\sqrt[3]{n-\sqrt{n^2+8}}$. We have gone in a perfect circle — and that is the point. **This expression and this equation are the same object in two costumes;** the problem was never asking us to compute, only to *recognize*. (Sanity check: $n = 10$ gives $x^3 + 6x = 20$, our Opening Challenge, with root 2.)

The equation-to-roots direction culminates in Cardano's formula; the roots-to-equation direction culminates in **Vieta's formulas**: for a monic cubic with roots r_1, r_2, r_3,

$$x^3 - (r_1{+}r_2{+}r_3)x^2 + (r_1r_2{+}r_2r_3{+}r_3r_1)x - r_1r_2r_3 = 0.$$

François Viète (1540–1603), one generation after *Ars Magna*, was the first to treat this direction as a theory in its own right, and the first to use letters systematically for both knowns and unknowns. If Cardano's book is modern algebra's birth certificate, Viète's notation is its alphabet.

The scandal of $\sqrt{-121}$

So far the formula has been odd but obedient. Now we feed it an innocent equation and watch it explode — productively.

WATCH THIS **#6 Bombelli, 1572**

Solve $x^3 = 15x + 4$.

Do it. By inspection $x = 4$ works: $64 = 60 + 4$. Factoring out $(x - 4)$ leaves $x^2 + 4x + 1$, so the three roots are

$$x = 4, \qquad x = -2 + \sqrt{3}, \qquad x = -2 - \sqrt{3}$$

— three honest real numbers. Now ask Cardano. With $p = -15, q = -4$:

$$\frac{q^2}{4} + \frac{p^3}{27} = 4 - 125 = -121,$$

and the formula insists $x = \sqrt[3]{2 + \sqrt{-121}} + \sqrt[3]{2 - \sqrt{-121}}$. A negative number under the square root — for an equation whose three roots we are *looking at*. Cardano knew of this case and essentially declared it beyond his art.

CURIOUS CARL • TRAPDOOR!

Carl: "A negative under the square root? That means *no real solutions* — like $x^2 + 1 = 0$. So the equation has none!"

The fall: but Carl, the equation's roots are 4, $-2 \pm \sqrt{3}$, sitting right there on the graph. The negative discriminant does *not* mean "no real roots" for a cubic — it means *three* real roots, and a formula that can only reach them by passing through $\sqrt{-1}$. The impossible number is a tunnel between real places, not a dead end.

Rafael Bombelli, an engineer who drained the marshes of the Val di Chiana and wrote his *Algebra* (1572) while a reclamation project stalled, refused to look away. His idea — which he himself called "a wild thought" (*un pensiero stravagante*) — was to treat $\sqrt{-1}$ as a number obeying the ordinary rules of arithmetic, and see whether the two impossible cube roots might cancel their impossible parts. He guessed the shape $\sqrt[3]{2 + 11\sqrt{-1}} = 2 + \sqrt{-1}$ — the same shape-guessing that tames an ordinary nested radical, now with $\sqrt{-1}$ in place of $\sqrt{2}$ — and checked it by cubing:

$$(2 + \sqrt{-1})^3 = 8 + 12\sqrt{-1} + 6(\sqrt{-1})^2 + (\sqrt{-1})^3 = 8 + 12\sqrt{-1} - 6 - \sqrt{-1} = 2 + 11\sqrt{-1}. \checkmark$$

Likewise $(2 - \sqrt{-1})^3 = 2 - 11\sqrt{-1}$, so

$$x = (2 + \sqrt{-1}) + (2 - \sqrt{-1}) = 4,$$

exactly the visible root. The impossible numbers entered, did their work, and left without a trace.

CURIOUS CARL • EUREKA

Carl watches two $\sqrt{-1}$'s dive into one end of the tunnel and a plain "4" stroll out the other. "You carried a real answer across on your back and then vanished before I could thank you. How am I supposed to call you *imaginary* now?"

This is the true birthplace of complex numbers. They were *not* invented to solve $x^2+1 = 0$; for centuries everyone was content to say that equation simply has no solutions. They were forced into existence by the *cubic*: by real equations with real roots whose only known formula routes through $\sqrt{-1}$. And Bombelli's predicament is not a fluke; it is the rule.

THE NAME FOR IT

Casus irreducibilis

For a depressed cubic $t^3 + pt + q = 0$ with rational coefficients, let $\Delta_0 = \frac{q^2}{4} + \frac{p^3}{27}$. If $\Delta_0 > 0$ there is one real root, reachable by Cardano with real radicals; if $\Delta_0 = 0$ the roots are repeated and rational in p, q; if $\Delta_0 < 0$ there are *three distinct real roots*, and if the cubic is irreducible over $\mathbb{Q}$ it is *provably impossible* to write any root using real radicals alone — every formula must detour through complex numbers. The Renaissance name for this last case, the "irreducible case," was an admission of defeat that turned out to be a theorem.

WHERE IT LEADS The impossibility in the third case was proved only in the nineteenth century, with *Galois theory*. The very cubic that names cos 20° is a casus irreducibilis; separately, its degree 3 (not a power of 2) is what makes a 60° angle impossible to trisect with straightedge and compass.

Sit with how strange the third case is: three real numbers, sitting visibly where the graph crosses the axis, and yet no expression built from rationals with $+, -, \times, \div$ and real nth roots can name them.

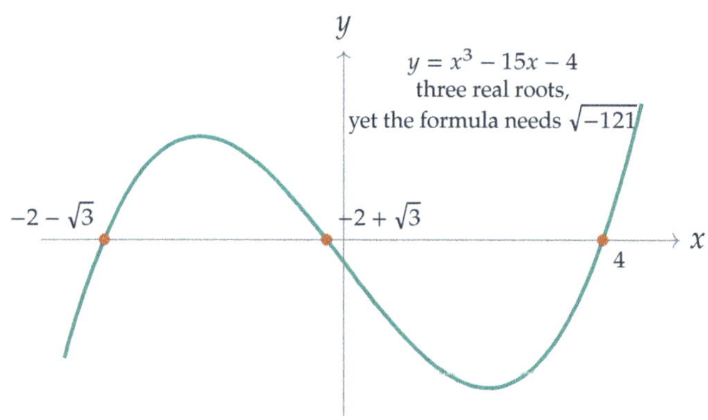

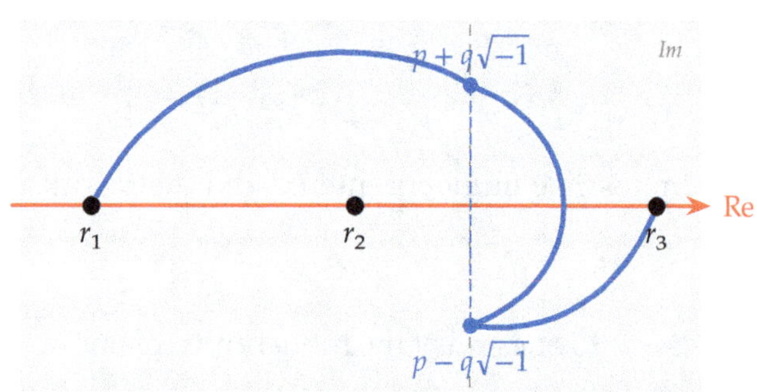

the only road back to a real answer runs through $\sqrt{-1}$

The three roots are real — yet Cardano's formula reaches them only by detouring through $\sqrt{-1}$ and back: the casus irreducibilis.

GOING FURTHER *optional • beyond AIME*

The escape hatch: trigonometry. If we allow tools beyond radicals, the three real roots of the casus irreducibilis have a clean closed form. For $t^3 + pt + q = 0$ with $\Delta_0 < 0$ (which forces $p < 0$),

$$t_k = 2\sqrt{-\tfrac{p}{3}}\,\cos\!\Big(\tfrac{1}{3}\arccos\!\big(\tfrac{3q}{2p}\sqrt{\tfrac{-3}{p}}\big) - \tfrac{2\pi k}{3}\Big), \qquad k = 0, 1, 2.$$

Why does trigonometry succeed where real radicals fail? Because the identity $\cos 3\theta = 4\cos^3\theta - 3\cos\theta$ *is* a depressed cubic: trisecting an angle means solving a cubic in $\cos\theta$, and for a 60° angle that cubic is a casus irreducibilis. It also explains why a 60° angle cannot be trisected with straightedge and compass: the cosine of the trisected angle has degree 3 over the rationals, while straightedge-and-compass constructions reach only lengths built from rationals by $+, -, \times, \div$ and *iterated square roots* — degrees that are powers of 2. You will meet the cosine machinery that powers this formula in the very next chapter.

Bombelli did more than rescue one equation. By writing down rules for adding and multiplying his "wild" quantities (what we now call the arithmetic of $a + bi$) he opened the door to the next chapter, where complex numbers stop being a detour and become a destination. Keep his lesson in hand: **complex numbers were legitimized not by philosophy but by the fact that they compute real answers correctly.**

Towers of radicals

Step back and look at the *shape* of Cardano's answer, $t = \sqrt[3]{-\frac{q}{2} + \sqrt{\Delta_0}} + \sqrt[3]{-\frac{q}{2} - \sqrt{\Delta_0}}$. It is a **nested radical**: a cube root wrapped around a square root. That is not an accident of the method. It is a portrait of the equation's inner structure. Solving the cubic takes two stages of root extraction, so its formula is a two-storey tower.

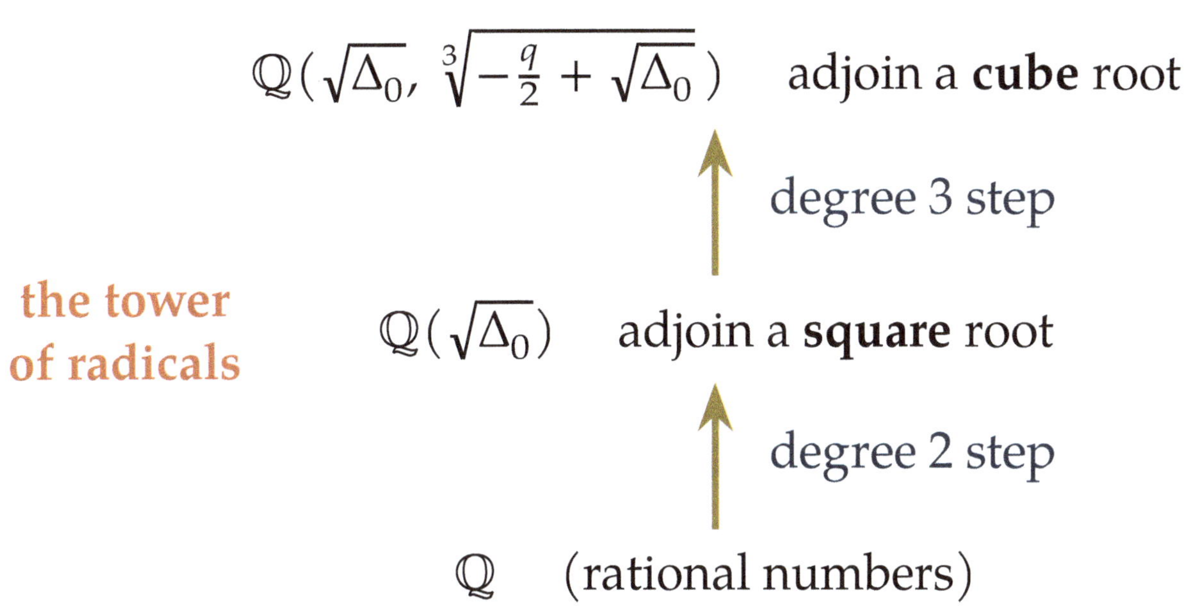

Every formula "by radicals" is such a tower: start from the rationals, repeatedly adjoin an nth root of something you already have, and reach the roots at the top floor. Ferrari's quartic solution is a three-storey tower (it reduces the quartic to a *resolvent cubic*, then stacks Cardano on top). And in this language, denesting acquires its true meaning: **to denest a radical is to discover that a tower has a hidden elevator** — that a number which appears to live on the second floor, like $\sqrt[3]{10+6\sqrt{3}}$, actually lives on the first, as $1+\sqrt{3}$.

After *Ars Magna* the program seemed clear: the quadratic needs one storey, the cubic two, the quartic three; surely the quintic needs four, and some heir of Cardano will find it. The greatest algebraists of three centuries (Euler and Lagrange among them) tried and failed. Lagrange (1770) found the deep reason all the old tricks worked: they secretly counted the ways the roots can be *permuted*, and his counting turned ominous at degree five.

Niels Henrik Abel (1802–1829), a Norwegian pastor's son so poor he published his great memoir at his own expense and trimmed it to six pages to afford the printing, proved in 1824 that the search was hopeless: *there is no general formula by radicals for the quintic*. No tower of any height reaches the top. He died of tuberculosis at 26, two days before a letter arrived offering him his first professorship. **Évariste Galois** (1811–1832) answered the deeper question (*which* equations are solvable by radicals, and why) by attaching to each equation the group of symmetries of its roots and reading the equation's fate from the structure of that group. A tower of radicals exists exactly when that symmetry group breaks into a chain of simple cyclic layers. The general quintic's group of 120 shufflings has an unbreakable core of 60; therefore no formula, ever. Galois wrote much of this the night before he died in a duel at twenty — a real duel, with pistols, unlike the mathematical one that opened this chapter.

Notice what Galois's answer is made of. Not a better formula — a change of *question*. Cardano asked "what is the root?"; Galois asked "what are the symmetries of the roots, viewed all together?" The subject born in 1545 with a formula reached maturity in 1832 by explaining why formulas run out.

GOING FURTHER *optional • beyond AIME*

Ramanujan, grandmaster of denesting. Nested radicals never stopped fascinating the masters. Srinivasa Ramanujan delighted in identities such as

$$\sqrt[3]{\sqrt[3]{2}-1}=\sqrt[3]{\tfrac{1}{9}}-\sqrt[3]{\tfrac{2}{9}}+\sqrt[3]{\tfrac{4}{9}},$$

a cube root denested into a three-term sum of cube roots (verify it by cubing the right side — a workout in $(a+b+c)^3$). Modern computer-algebra systems denest radicals with algorithms whose correctness proofs run on Galois theory: the shape-guessing ansatz of this chapter, industrialized. The circle closes — the technique that fell out of Cardano's formula as a byproduct became, four centuries later, a test case for the theory that explained the formula.

What to try when you're stuck

When a problem hands you a cubic, or a tower of cube-and-square roots:

1. **Depress first.** Substitute $x=t-\frac{a}{3}$ to kill the square term; now p and q are all that matter.

2. **Guess the shape.** Write the unknown as $t = u + v$ and *spend* the extra freedom ($uv = -\frac{p}{3}$) to collapse the cubic into a quadratic in u^3.

3. **Read a nested radical as a root.** Given $\sqrt[3]{A} + \sqrt[3]{B}$, form $s = A + B$, $P = \sqrt[3]{AB}$, and write the cubic $x^3 - 3Px - s = 0$; then hunt a rational root.

4. **Rational-root test, then uniqueness.** If the cubic is increasing ($P \leq 0$), a rational root is the *only* real root, so the radical equals it.

5. **Mind the sign of a cube root.** $\sqrt[3]{\text{negative}}$ is negative — denest to $m - n\sqrt{d}$, not $n\sqrt{d} - m$.

6. **Read the discriminant Δ_0.** $\Delta_0 < 0$ means *three* real roots and an unavoidable detour through $\sqrt{-1}$ — reach for trigonometry, not more radicals.

THE MOVE

One-page recap. The cubic yields to a single audacious move: **guess the shape** $t = u + v$, spend the spare freedom to make the middle term vanish, and a cube becomes a quadratic — **Cardano's formula**, a square root nested in two cube roots. The formula runs both ways: **equation → roots** (manufacture u, v) and **roots → equation** (turn $\sqrt[3]{A} + \sqrt[3]{B}$ into a clean cubic and read off its rational root). **Denesting** is the same shape-guessing one storey down. When the discriminant goes negative — three real roots you can see — the formula must pass through $\sqrt{-1}$: the **casus irreducibilis**, where the imaginary numbers were forced into being. Guess the shape; travel both directions; when radicals run out, change the question.

SEEN BEFORE?

The "spend a spare degree of freedom to make a term vanish" gambit is the same instinct as the *clever substitution* that collapses a hard sum, and the guess-the-shape ansatz echoes every *undetermined-coefficients* argument. Most of all, the impossible numbers that cancel here become the star of **Chapter 13** (*Tools From the Imaginary Box*), where $a + bi$ gets a geometry — addition as translation, multiplication as rotation — and the $\cos 3\theta$ identity behind the trig escape hatch returns as de Moivre's theorem. The "symmetry of the roots" that decides solvability is the same counting-under-symmetry idea that fills **Chapter 16** (*Counting Past the Mirror*).

The practice ladder

THE PRACTICE LADDER *climb at your own pace*

LAUNCH *The formula, forward. Nudges below.*

1. Solve $x^3 + 3x - 14 = 0$ by Cardano's formula, then denest to show the root is an integer.

2. Solve $x^3 - 6x - 40 = 0$ by Cardano's formula and denest. (You should meet $\sqrt{392} = 14\sqrt{2}$.)

CORE BUILDER *Master both directions.*

1. Solve $x^3 - 6x^2 + 6x - 2 = 0$ exactly. First depress, then apply Cardano; the cube roots will *not* denest.

2. Simplify $\sqrt[3]{2+\sqrt{5}} + \sqrt[3]{2-\sqrt{5}}$.

3. Simplify $\sqrt[3]{5+2\sqrt{13}} + \sqrt[3]{5-2\sqrt{13}}$.

STRETCH *Transfer to unfamiliar ground.*

1. Prove that $\sqrt[3]{\sqrt{5}+2} - \sqrt[3]{\sqrt{5}-2} = 1$. (Careful: this one is a *difference* — adapt the cubing identity.)

2. Simplify $\sqrt[3]{26+15\sqrt{3}} + \sqrt[3]{26-15\sqrt{3}}$, and explain how you know both real cube roots are positive before you begin.

THE FLASH *No clock — each cracks open in one flash of insight, once you find the angle.*

1. Find all positive integers n for which $\sqrt[3]{n+\sqrt{n^2+8}} + \sqrt[3]{n-\sqrt{n^2+8}}$ is an integer. (Hint: if the value is k, then $k^3 + 6k = 2n$; now think parity.)

PROOF LAB *Write it clean, in full sentences.*

1. Find the monic cubic with integer coefficients whose root is $1+\sqrt[3]{2}$ (its *minimal polynomial*).

2. The equation $x^3 - 7x + 6 = 0$ factors nicely; find its three roots by inspection. Then compute $\Delta_0 = \frac{q^2}{4} + \frac{p^3}{27}$ and write down (do not simplify) what Cardano's formula asserts about the root $x = 1$ — your own Bombelli-style identity.

3. Show that if a depressed cubic $t^3 + pt + q$ with rational p, q has a rational root r, then $-\frac{q}{2} + \sqrt{\Delta_0}$ is the cube of a number of the form $m + n\sqrt{d}$ with rational m, n. (This justifies "rational root $\Longleftrightarrow$ denestable formula.")

4. Use $\cos 3\theta = 4\cos^3\theta - 3\cos\theta$ to write $\cos 20^\circ$ as a root of a cubic; verify the cubic falls into the casus irreducibilis; and explain in a paragraph why this shows a 60° angle cannot be trisected with straightedge and compass — recalling that such constructions reach exactly the lengths built from the rationals by $+, -, \times, \div$ and *iterated square roots*.

REFLECTION *No nudge. Write a few sentences.*

1. Complex numbers were *not* forced into being by $x^2 + 1 = 0$, but by the cubic. In your own words, what is it about a formula being the *only* road to a *real* answer that makes a number impossible to dismiss?

NUDGES

Launch. (1) $\Delta_0 = 49 + 1 = 50$; $x = \sqrt[3]{7+5\sqrt{2}} + \sqrt[3]{7-5\sqrt{2}} = (1+\sqrt{2}) + (1-\sqrt{2}) = 2$. (2) $x = \sqrt[3]{20+14\sqrt{2}} +$

$\sqrt[3]{20-14\sqrt{2}} = (2+\sqrt{2})+(2-\sqrt{2}) = 4$. **Core.** (3) Depress with $x = t+2$ to get $t^3-6t-6 = 0$; the root is $x = 2+\sqrt[3]{2}+\sqrt[3]{4}$. (4) $s = 4, AB = -1, P = -1$: $x^3+3x-4 = 0$, so $x = 1$. (5) $AB = -27, P = -3$: $x^3+9x-10 = 0$, so $x = 1$. **Stretch.** (6) Set $x = a-b$ with $a = \sqrt[3]{\sqrt{5}+2}, b = \sqrt[3]{\sqrt{5}-2}$: then $ab = \sqrt[3]{5-4} = 1$ and $a^3-b^3 = 4$, so $x^3+3x-4 = 0$ and $x = 1$. (7) $(2\pm\sqrt{3})^3 = 26\pm 15\sqrt{3}$; both cube roots are positive since $15\sqrt{3} < 26$; sum $= 4$. **The Flash.** (8) If the value is k then $2n = k^3+6k$; k must be even (odd k makes k^3+6k odd), so $k = 2m$ gives $n = 4m^3+6m = 10, 44, 126, \ldots$ **Proof.** (9) With $x = 1+\sqrt[3]{2}$, we have $x-1 = \sqrt[3]{2}$, so $(x-1)^3 = 2$ gives $x^3-3x^2+3x-3 = 0$; this monic cubic has no rational root, hence is irreducible over $\mathbb{Q}$ and is the minimal polynomial. (10) By inspection $x^3-7x+6 = (x-1)(x-2)(x+3)$, so the roots are $1, 2, -3$. With $p = -7, q = 6, \Delta_0 = \frac{q^2}{4}+\frac{p^3}{27} = 9-\frac{343}{27} = -\frac{100}{27} < 0$ — the casus irreducibilis — yet Cardano's formula asserts $1 = \sqrt[3]{-3+\frac{10}{9}\sqrt{-3}}+\sqrt[3]{-3-\frac{10}{9}\sqrt{-3}}$, a real root built from conjugate complex cube roots. (11) A rational root r lets you factor $t^3+pt+q = (t-r)(t^2+rt+\beta)$ with $\beta = r^2+p$ (and $q = -r\beta$), so the remaining roots — and all the structure of Δ_0 — live in the quadratic field $\mathbb{Q}(\sqrt{r^2-4\beta})$. Since $\sqrt{\Delta_0}$ lies in that same field, Cardano's $u^3 = -\frac{q}{2}+\sqrt{\Delta_0}$ admits a cube root of the form $m+n\sqrt{d}$ with rational m, n; adding u to its conjugate then recovers r with the nested radicals denested. This is why a rational root is exactly what collapses the formula to a clean $m+n\sqrt{d}$. (12) Since $3\theta = 60^\circ$ and $\cos 60^\circ = \frac{1}{2}$, the identity gives $4c^3-3c = \frac{1}{2}$, i.e. $8c^3-6c-1 = 0$ for $c = \cos 20^\circ$; the rational-root test shows this is irreducible over $\mathbb{Q}$, and its three real roots ($\Delta_0 < 0$) place it in the casus irreducibilis. Thus $\cos 20^\circ$ has degree 3 over $\mathbb{Q}$, while every straightedge-and-compass length lies in a tower of quadratic extensions and so has degree a power of 2. As 3 is not a power of 2, $\cos 20^\circ$ is not constructible, and therefore a 60° angle cannot be trisected.

CHAPTER THIRTEEN

13

Tools From the Imaginary Box

Mathematicians once called $\sqrt{-1}$ impossible and used it anyway, holding their noses. Reach into that box of "imaginary" tools and real problems start falling open.

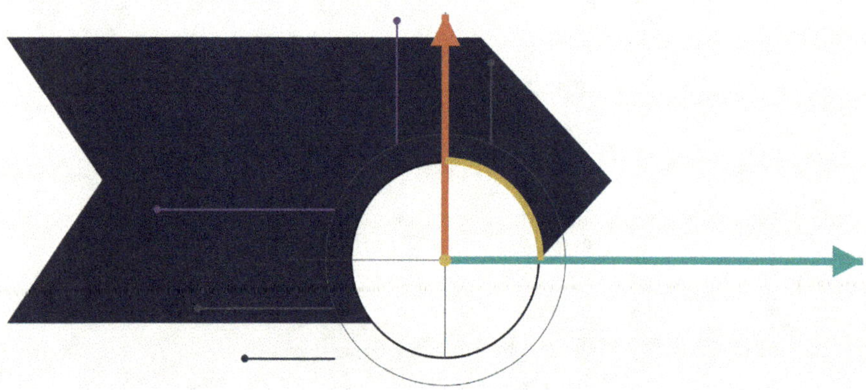

Complex numbers: to turn, multiply.

Here is a fact that looks like it needs trigonometry — and doesn't.

WONDER THIS

Mark five equally spaced points around a circle — the corners of a regular pentagon, with one corner at the rightmost point $(1, 0)$. Treat each as an arrow from the center.

Add all five arrows together. What do you get? The answer is exactly zero — the arrows perfectly cancel. **Why must they?**

CURIOUS CARL • FIRST GUESS

Carl: "Add $\cos 0^\circ + \cos 72^\circ + \cdots + \cos 288^\circ$? Fine. Pair the angles, collapse each pair with the sum-to-product identity $\cos A + \cos B = 2\cos\frac{A+B}{2}\cos\frac{A-B}{2}$, and grind through what's left. A page or two of algebra, but it closes."

The catch: your method genuinely works, Carl — it just costs pages of ugly irrationals. But treat the five points as fifth roots of unity, and multiplying their whole sum by one root just *rotates it onto itself*, forcing the total to 0 with no identities at all.

You could set up five pairs of sines and cosines and grind through the sum, fighting messy values like $\cos 72^\circ$. But there's a way to see the cancellation with almost no computation, and it comes from reaching into a box of tools mathematicians once distrusted: the **complex numbers**. Treat the five points as complex numbers and they become the five solutions of one clean equation, and their sum emerges from the equation's structure.

A *complex number* $a + bi$, where $i^2 = -1$, is best pictured not as an exotic quantity but as a *point in the plane* — or an arrow from the origin to that point. The number 1 points right; i points up; -1 points left. Addition is just adding arrows tip-to-tail, exactly like vectors. But complex numbers can also be *multiplied*, and that's where the magic lives: multiplying by a complex number *rotates and scales* the plane. This fusion (arrows you can both add, like geometry, and multiply, like algebra) is what fills the toolbox. In Chapter 12 these "impossible" numbers were forced into being by Cardano's cubic and then politely vanished from the answer; here they earn a permanent home (a geometry of their own) and stop being a trick you hold your nose to use.

THE MOVE

A complex number has two faces. In **rectangular form** $a+bi$, it's a point (a, b) — good for adding. In **polar form** $r(\cos\theta + i\sin\theta)$, it's an arrow of length r at angle θ — good for multiplying, because *multiplying multiplies the lengths and adds the angles*. Switching between the two faces, using each where it's strong, is the central skill. Geometry and algebra become one.

That "multiply lengths, add angles" rule is the key that unlocks the whole box. It means raising a complex number to a power just multiplies its angle, so questions about powers, rotations, and regularly-spaced points all become easy. The five pentagon corners are about to become the five fifth-roots of the number 1.

Roots of unity: the equation behind the pentagon

The pentagon points are the complex solutions of $z^5 = 1$: the *fifth roots of unity*. Let's see why, and why they cancel.

WATCH THIS **#1 The nth roots of unity**

Setup. We want all complex z with $z^n = 1$. Writing z in polar form as an arrow of length r at angle θ, raising to the nth power gives an arrow of length r^n at angle $n\theta$. For this to equal 1 (length 1, angle 0 or any full turn), we need $r = 1$ and $n\theta$ a whole number of turns.

The roots. So the solutions are arrows of length 1 at angles $\theta = \frac{2\pi k}{n}$ for $k = 0, 1, \ldots, n-1$ — exactly n points, equally spaced around the unit circle, forming a regular n-gon.

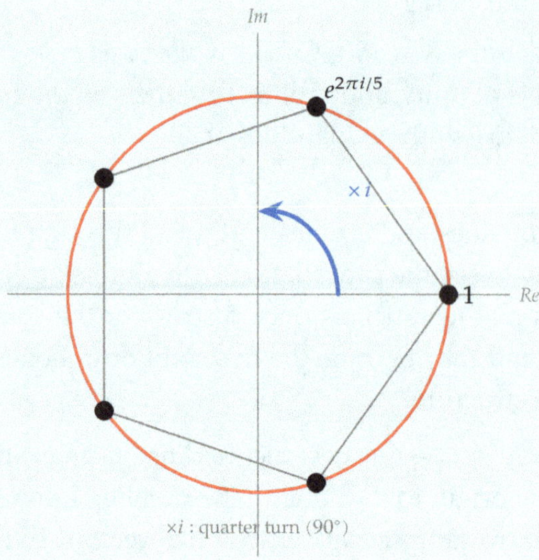

Why they sum to zero. The roots are $1, \zeta, \zeta^2, \ldots, \zeta^{n-1}$ where $\zeta = \cos\frac{2\pi}{n} + i\sin\frac{2\pi}{n}$ is the first one past 1. Their sum $S = 1 + \zeta + \cdots + \zeta^{n-1}$ is a geometric series. Multiply it by ζ: this just *rotates each root to the next one*, permuting them, so $\zeta S = S$. Then $(\zeta - 1)S = 0$, and since $\zeta \neq 1$, we must have $S = 0$. ∎

Trigger Anything involving regularly spaced points on a circle, powers that cycle, rotational symmetry, or sums of equally spaced angles.

Do first Reframe the points as roots of unity $z^n = 1$; use that multiplying by ζ rotates the configuration, so symmetric sums collapse.

Trapdoor Forgetting the root $k = 0$ (the number 1 itself) or miscounting: there are exactly n roots, one for each k from 0 to $n - 1$.

Upgrade The "multiply by ζ to rotate" trick evaluates many symmetric sums; the roots-of-unity filter (later) extracts every dth term of any sum.

So the pentagon's arrows cancel for a reason far cleaner than trigonometry: they're the roots of $z^5 = 1$, and rotating the whole set by one notch leaves it unchanged, which forces their sum to be fixed by a rotation, and the only point fixed by a nontrivial rotation is the center, zero. *Symmetry, not computation, did the work.*

THE TURN *from grind to insight*

The obvious attack is to write each arrow as a (cos, sin) pair and add the five sines and five cosines by hand, wrestling with ugly values like cos 72° and half-remembered angle-sum identities until the mess somehow collapses to zero. **The creative turn** is to treat each point as a complex number, so the five arrows become $1, \zeta, \zeta^2, \zeta^3, \zeta^4$ — and *multiplying the whole sum by* *ζ merely rotates the set onto itself*, forcing $S = \zeta S$, hence $S = 0$. It wins because rotation is just multiplication here: the symmetry that was invisible in trig becomes a one-line algebraic fact, and not a single cosine was ever evaluated.

De Moivre: powers become rotations

The "add the angles" rule, applied to powers, is so useful it has a name.

WATCH THIS #2 De Moivre's theorem

Claim. $(\cos\theta + i\sin\theta)^n = \cos(n\theta) + i\sin(n\theta)$.

Why. A complex number of length 1 at angle θ, raised to the nth power, has length $1^n = 1$ and angle $n\theta$ — the multiplication rule applied n times. So the nth power is the arrow of length 1 at angle $n\theta$, which is $\cos(n\theta) + i\sin(n\theta)$.

A bonus: trig identities for free. Expand the left side for $n = 2$ using $i^2 = -1$:

$$(\cos\theta + i\sin\theta)^2 = \cos^2\theta - \sin^2\theta + i\,(2\sin\theta\cos\theta).$$

Matching real and imaginary parts to $\cos 2\theta + i\sin 2\theta$ instantly gives the double-angle formulas $\cos 2\theta = \cos^2\theta - \sin^2\theta$ and $\sin 2\theta = 2\sin\theta\cos\theta$ — no memorization required. ∎

The trigonometric identities you once memorized are just the real and imaginary parts of a single complex power. De Moivre's theorem turns every "angle $n\theta$" question into an algebra problem about expanding a power, and the binomial theorem from Chapter 7 does the expanding. Whenever you see $\cos(n\theta)$ or $\sin(n\theta)$, think: *that's the real or imaginary part of* $(\cos\theta + i\sin\theta)^n$.

△ TRAPDOOR

Angles in complex numbers are only determined *up to full turns* of 2π. When taking nth roots, you get n different answers precisely because adding $2\pi k$ to the angle before dividing by n lands on n distinct results. Forgetting this gives you one root instead of all n. Always sweep $k = 0, 1, \dots, n-1$ to capture every root.

CURIOUS CARL • TRAPDOOR!

Carl wants to spin a point a quarter-turn, so he multiplies by $z = 1 + i$. "There — rotated!" But his point has drifted farther from the origin, and he can't see why.

The stray arrow protests: *"You didn't just turn me, you stretched me —* $|1+i| = \sqrt{2}$, *so you scaled my length by* $\sqrt{2}$ *on the way round. A clean turn needs* $|z| = 1$." Careful, Carl — multiplying rotates *and* scales; only when the length is 1 (like i itself) is it a pure rotation.

Powers that obey a recurrence

De Moivre turns $\cos n\theta$ into a power you expand. There is an even lazier route: a two-line *recurrence* that grinds out $\cos n\theta$ — and much more — without expanding anything at all.

THE MOVE

Conjugate powers run on a recurrence. For any number z, the sum $s_n = z^n + \bar{z}^n$ is real, and because z and $\bar{z}$ are the two roots of $t^2 - (z+\bar{z})\,t + z\bar{z} = 0$, the sequence obeys

$$s_{n+1} = (z+\bar{z})\,s_n - (z\bar{z})\,s_{n-1}, \qquad s_0 = 2,\ s_1 = z + \bar{z}.$$

The whole sequence is built from its first two terms by addition and multiplication — no binomial expansion, no trig table. A high power of z collapses onto a short recurrence.

WATCH THIS #3 One recurrence, two faces

Face 1: cosines (on the unit circle). Take $z = \cos\theta + i\sin\theta$, so $\bar{z} = \cos\theta - i\sin\theta$, $z\bar{z} = 1$, and $z + \bar{z} = 2\cos\theta$. Then $s_n = 2\cos n\theta$ satisfies

$$s_{n+1} = 2\cos\theta\, s_n - s_{n-1}, \qquad s_0 = 2,\ s_1 = 2\cos\theta.$$

Turn the crank: $s_2 = (2\cos\theta)^2 - 2 = 2(2\cos^2\theta - 1) = 2\cos 2\theta$, then $s_3 = 2\cos 3\theta$, and so on. Every $2\cos n\theta$ is a fixed integer polynomial in $2\cos\theta$ — the *Chebyshev polynomials* — so a power of a cosine is really a *sum*, never a product to expand.

Face 2: surds (off the circle). Take the conjugates $\alpha = 2 + \sqrt{3}$ and $\beta = 2 - \sqrt{3}$, with $\alpha + \beta = 4$ and $\alpha\beta = 1$. Then $s_n = \alpha^n + \beta^n$ satisfies

$$s_{n+1} = 4s_n - s_{n-1}, \qquad s_0 = 2,\ s_1 = 4 \implies 2, 4, 14, 52, 194, 724, \ldots$$

Every term is an integer, though each is a sum of two irrational powers. ■

Face 2 hides a small miracle. Since $0 < \beta = 2 - \sqrt{3} < 1$, the term β^n is a vanishing crumb, so $\alpha^n = s_n - \beta^n$ sits just below the integer s_n. That pins down a quantity you would never want to expand:

$$\lceil (2+\sqrt{3})^n \rceil = s_n, \qquad \text{e.g. } (2+\sqrt{3})^4 = 194 - \beta^4 = 193.994\ldots, \text{ so } \lceil (2+\sqrt{3})^4 \rceil = 194.$$

The ceiling of a runaway irrational power is an ordinary integer, read straight off a recurrence — the escaping partner α^n is fenced in by its shrinking twin. *Don't expand the power; find the recurrence it already obeys.*

SEEN BEFORE?

This is the linear recurrence of **Chapter 5** arriving from the other side: there we *built* a recurrence to count; here we notice that $z^n + \bar{z}^n$ *already* satisfies one, simply because z and $\bar{z}$ are the two roots of a single quadratic — and a root pair's sum and product are exactly the recurrence's coefficients. Face 2 is also **Chapter 10**'s "giants collapse" in miniature: an astronomical power is trapped between consecutive integers by its tiny conjugate. And the cosine sum a few pages on is this same $z^n + \bar{z}^n$ move with θ a pentagon angle.

Two gems from the box: a sine product and the golden pentagon

Before the showpiece, two results that look like they belong to trigonometry or geometry but fall out of $z^n - 1$ with almost no effort. They show how varied the box's tools are.

WATCH THIS **#4 A product of sines equals n**

Claim. $\prod_{k=1}^{n-1} 2\sin\frac{k\pi}{n} = n$. (For $n = 6$: $2\sin 30^\circ \cdot 2\sin 60^\circ \cdots = 6$.)

Factor $z^n - 1$ by its roots. Since the roots of $z^n - 1$ are ζ^k for $k = 0, \dots, n-1$, and the $k = 0$ root is $z = 1$,

$$z^n - 1 = (z-1)\prod_{k=1}^{n-1}(z - \zeta^k) \quad\Longrightarrow\quad \prod_{k=1}^{n-1}(z-\zeta^k) = \frac{z^n - 1}{z - 1} = 1 + z + \cdots + z^{n-1}.$$

Set $z = 1$: the right side is n, so $\prod_{k=1}^{n-1}(1 - \zeta^k) = n$.

Turn each factor into a sine. The chord from 1 to ζ^k has length $|1 - \zeta^k| = 2\sin\frac{k\pi}{n}$ (an isosceles triangle of two unit radii at angle $\frac{2\pi k}{n}$). Taking absolute values, $\prod_{k=1}^{n-1} 2\sin\frac{k\pi}{n} = |n| = n$. **The product of all those sines is just the integer n** — the distances from one polygon vertex to all the others multiply to exactly n. ∎

WATCH THIS **#5 $\cos 36^\circ$ and the golden ratio**

Claim. $\cos 36^\circ = \dfrac{1+\sqrt{5}}{4} \approx 0.809$, exactly half the golden ratio.

A pentagon identity from the root sum. The five fifth-roots of unity sum to zero, so their real parts do too: $1 + 2\cos 72^\circ + 2\cos 144^\circ = 0$ (pairing each root with its conjugate). Now $\cos 144^\circ = -\cos 36^\circ$, so $1 + 2\cos 72^\circ - 2\cos 36^\circ = 0$, which rearranges to the clean relation

$$\cos 36^\circ - \cos 72^\circ = \tfrac{1}{2}.$$

Solve the quadratic. Write $c = \cos 36^\circ$. The double-angle formula gives $\cos 72^\circ = 2c^2 - 1$, so $c - (2c^2 - 1) = \frac{1}{2}$, that is

$$4c^2 - 2c - 1 = 0 \quad\Longrightarrow\quad c = \frac{2 + \sqrt{4+16}}{8} = \frac{1+\sqrt{5}}{4}.$$

(We take the positive root, since $\cos 36^\circ > 0$.) So $\cos 36^\circ = \dfrac{1+\sqrt{5}}{4} \approx 0.809$. **This is exactly half the golden ratio: $\cos 36^\circ = \frac{\varphi}{2}$, where $\varphi = \frac{1+\sqrt{5}}{2}$.** ∎

The golden ratio $\varphi = \frac{1+\sqrt{5}}{2}$ is hiding inside the regular pentagon precisely because the pentagon's geometry *is* the fifth roots of unity, and their algebra is governed by $z^5 = 1$. The diagonal-to-side ratio of a regular pentagon is φ; the appearance of $\sqrt{5}$ in $\cos 36^\circ$ is the same fact wearing a trigonometric costume. *One equation, $z^5 = 1$, secretly runs the whole pentagon.*

The showpiece: the roots-of-unity filter

Here's the box's most surprising tool: using complex numbers to extract hidden patterns from ordinary sums of integers.

WONDER THIS

In a row of Pascal's triangle, add up only *every third* entry: $\binom{n}{0} + \binom{n}{3} + \binom{n}{6} + \cdots$

Is there a clean formula for this partial row sum? The binomial theorem gives the *whole* row easily — but how do you isolate just the multiples-of-three?

The trouble is sieving out two of every three terms. Real algebra can't easily do it — but the roots of unity can, because they have a magical averaging property.

WATCH THIS **#6 Sieving with roots of unity**

The magic property. Let $\omega = e^{2\pi i/3}$ be a primitive cube root of unity, so $1 + \omega + \omega^2 = 0$. For any integer j, the sum $1 + \omega^j + \omega^{2j}$ equals 3 if j is a multiple of 3, and 0 otherwise. (When $3 \mid j$ each term is 1; otherwise the three terms are the three cube roots of unity, summing to 0.) This is a *switch* that turns on only for multiples of three.

Apply it to the binomial theorem. Evaluate $(1 + x)^n$ at $x = 1, \omega, \omega^2$ and add:

$$(1+1)^n + (1+\omega)^n + (1+\omega^2)^n = \sum_{k=0}^{n} \binom{n}{k} (1 + \omega^k + \omega^{2k}).$$

By the magic property, the bracket is 3 when $3 \mid k$ and 0 otherwise — so the right side is 3 times exactly the sum we want!

The result. Dividing by 3,

$$\binom{n}{0} + \binom{n}{3} + \binom{n}{6} + \cdots = \frac{2^n + (1+\omega)^n + (1+\omega^2)^n}{3} = \frac{2^n + 2\cos\frac{n\pi}{3}}{3}.$$

For $n = 6$: $\frac{64+2\cos 2\pi}{3} = \frac{64+2}{3} = 22$, which checks against $\binom{6}{0} + \binom{6}{3} + \binom{6}{6} = 1 + 20 + 1 = 22$. ■

The general tool. Nothing was special about 3. To sieve *every dth term*, average $(1 + x)^n$ over the dth roots of unity:

$$\sum_{d \mid k} \binom{n}{k} = \frac{1}{d} \sum_{j=0}^{d-1} (1 + \omega^j)^n, \qquad \omega = e^{2\pi i/d}.$$

The same on/off switch $\frac{1}{d}\sum_j \omega^{jk}$ keeps the terms with $d \mid k$ and kills the rest. For instance, every *fourth* term of row 8 is $\binom{8}{0} + \binom{8}{4} + \binom{8}{8} = 1 + 70 + 1 = 72$, which the $d = 4$ filter reproduces. This is the showpiece because of what happens: to extract a real pattern, we detour through the complex plane, use the cancellation of roots of unity as a switch, and return with a clean real formula, the imaginary parts vanishing exactly as they must. *The complex numbers were never the destination — they were the bridge, and the answer was real all along.*

Roots of unity are a switchboard. Don't picture the dth roots as mere points on a circle; picture them as *switches*, one per residue class, that the averaging turns on and off:

$$\frac{1}{d}\left(1+\omega^k+\omega^{2k}+\cdots+\omega^{(d-1)k}\right)=\begin{cases}1, & d \mid k,\\ 0, & d \nmid k.\end{cases}$$

Flip the right combination of switches and any residue class you like lights up while the rest go dark. This is not a coincidence of algebra — it is the smallest *discrete Fourier transform*: averaging a signal against rotating phases ω^{jk} picks out one frequency and cancels the others, which is exactly how your phone separates the tones in a sound or the colors in an image. *Don't fight the combinatorics head-on — move into the complex plane and let the switches sort the terms.*

TRY TO BREAK IT

The nth roots of unity sum to 0 — but the proof needed $\zeta \neq 1$, so it quietly assumed $n \geq 2$. Try it for $n = 1$: the equation $z^1 = 1$ has the single root $\{1\}$, and its "sum" is 1, not 0. Trace the "multiply by ζ and rotate" argument in this case and pinpoint the exact step where it breaks — why can't you conclude $S = 0$ when there's only one root to rotate?

SEEN BEFORE?

This filter sieves binomial coefficients by their residue mod d — a fusion of the binomial theorem (**Chapter 7**) with the modular thinking of **Chapter 9**. The "$1+\omega+\omega^2=0$" fact is the roots-of-unity sum from this chapter's opener, now working as a switch, and expanding $(1+\omega)^n$ uses de Moivre. The sine-product gem is the same factorization of $z^n - 1$ that powers the filter.

Counting by residue: how many subsets have a given sum?

The filter sieved a *known* sum: a row of Pascal's triangle. The same switch does something bolder: it *counts* combinatorial objects sorted by a residue, a job with no elementary shortcut.

WONDER THIS

Of the $2^6 = 64$ subsets of $\{1,2,3,4,5,6\}$ (counting the empty set, sum 0), how many have a sum *divisible by* 3?

Listing all 64 and checking each is grim. Can you count them without enumerating?

Pack all the subsets into a single polynomial, then let the roots of unity read off just the multiples of three.

WATCH THIS #7 Subsets with sum divisible by 3

Pack the subsets into a product. Each element j is either out (contribute 1) or in (contribute x^j). Multiplying the choices,

$$P(x) = \prod_{j=1}^{6}(1 + x^j) = \sum_{S \subseteq \{1,\dots,6\}} x^{\text{sum}(S)},$$

so the coefficient of x^s counts the subsets with sum s. We want the total of all coefficients whose exponent is divisible by 3 — exactly what the filter extracts.

Average over the cube roots of unity. With $\omega = e^{2\pi i/3}$,

$$\#\{\text{subsets with } 3 \mid \text{sum}\} = \frac{P(1) + P(\omega) + P(\omega^2)}{3}.$$

Here $P(1) = 2^6 = 64$. For $P(\omega)$, group the six factors by j mod 3: the exponents $1, 2, 3, 4, 5, 6$ give

$$(1 + \omega)(1 + \omega^2)(1 + 1)(1 + \omega)(1 + \omega^2)(1 + 1).$$

Since $(1 + \omega)(1 + \omega^2) = 1 + \omega + \omega^2 + \omega^3 = 1$, this collapses to $1 \cdot 1 \cdot 2 \cdot 2 = 4$; the same count gives $P(\omega^2) = 4$.

The count.

$$\frac{64 + 4 + 4}{3} = \frac{72}{3} = 24.$$

Exactly 24 of the 64 subsets have sum divisible by 3 — found without listing a single one. ■

THE MOVE

Counting by residue. To count objects whose "size" (sum, weight, length) is $\equiv r \pmod{d}$, build the generating polynomial $P(x)$ that records each object as x^{size}, then average P over the dth roots of unity (weighting by ω^{-rj} to select residue r). The complex evaluations collapse the whole polynomial into one integer count. It is the roots-of-unity filter pointed at *counting*, not at a fixed sum.

SEEN BEFORE?

The polynomial $P(x) = \prod_j (1 + x^j)$ is a *generating function* (**Chapter 8**): a product encodes independent in/out choices, and a coefficient counts configurations. Here complex numbers read a single residue class out of it — generating functions supply the object, roots of unity supply the sieve.

Payoff: a cosine sum with no trig

To close, the kind of sum that looks like it needs a trig table but yields to one idea.

WONDER THIS

Add up the cosines at the five pentagon angles: $\cos 0^\circ + \cos 72^\circ + \cos 144^\circ + \cos 216^\circ + \cos 288^\circ$.

What's the total? Those individual cosines are ugly irrational numbers — but the sum is beautifully simple.

Don't touch a single cosine value. The five fifth-roots of unity sum to zero — we proved it in the opener. Each root is $\cos \frac{2\pi k}{5} + i \sin \frac{2\pi k}{5}$, so the *real parts* of the roots are exactly our five cosines. If the whole complex sum is zero, then its real part is zero too:

$$\cos 0^\circ + \cos 72^\circ + \cos 144^\circ + \cos 216^\circ + \cos 288^\circ = \mathrm{Re}(0) = 0.$$

The five cosines sum to zero — because they're the shadows of arrows that already cancel. No trig identities, no irrational arithmetic; just the observation that the real part of zero is zero. *Sometimes the fastest route between two real numbers runs through the imaginary plane.*

A different box: the Gaussian integers

Complex numbers do more than rotate. They form a whole *number system*, the *Gaussian integers* $\mathbb{Z}[i] = \{a + bi : a, b \in \mathbb{Z}\}$, in which stubborn whole-number questions factor open. The key is the *norm* $N(a + bi) = a^2 + b^2 = |a + bi|^2$: a sum of two squares is a squared length in disguise.

WONDER THIS

$5 = 1^2 + 2^2$ and $13 = 2^2 + 3^2$ are each a sum of two squares. Must their product 65 be a sum of two squares too?

And is there a reason the answer is always yes, for any two such numbers?

THE MOVE

Sums of two squares are norms, and norms multiply. Because $a^2+b^2 = |a+bi|^2$ and $|zw| = |z|\,|w|$, the product of two sums of two squares is again a sum of two squares. The whole bookkeeping of "which numbers are sums of two squares" is really the geometry of lengths under complex multiplication.

WATCH THIS #8 The product of two sums of two squares

One line from $|zw| = |z|\,|w|$. Let $z = a + bi$ and $w = c + di$, so $|z|^2 = a^2 + b^2$ and $|w|^2 = c^2 + d^2$. Then

$$(a^2 + b^2)(c^2 + d^2) = |z|^2|w|^2 = |zw|^2.$$

Now $zw = (ac - bd) + (ad + bc)i$, whose norm is $(ac - bd)^2 + (ad + bc)^2$. Equating the two,

$$(a^2 + b^2)(c^2 + d^2) = (ac - bd)^2 + (ad + bc)^2,$$

the *Brahmagupta–Fibonacci identity*. For $5 \cdot 13$: take $z = 1 + 2i$, $w = 2 + 3i$, so $zw = -4 + 7i$ and $65 = 4^2 + 7^2$. (Conjugating w to $2 - 3i$ gives $zw = 8 + i$, the other form $65 = 8^2 + 1^2$.) ■

The norm also decides which *primes* are sums of two squares. A prime $p \equiv 1 \pmod 4$ *splits* in the Gaussian integers, factoring as $p = (a + bi)(a - bi) = a^2 + b^2$ (for instance $13 = (2 + 3i)(2 - 3i)$), so it is a sum of two squares. A prime $p \equiv 3 \pmod 4$ never is, and Chapter 9 already gave the reason.

SEEN BEFORE?

A sum of two squares is $\equiv 0, 1$, or $2 \pmod 4$, *never* 3 — the impossibility filter from **Chapter 9**. So no prime $p \equiv 3 \pmod 4$ is a sum of two squares, while every prime $p \equiv 1 \pmod 4$ is (Fermat's two-square theorem). The Gaussian integers explain *why*: primes $\equiv 1$ factor, primes $\equiv 3$ stay whole.

GOING FURTHER *optional • beyond AIME*

Why must a prime $p \equiv 1 \pmod 4$ split? Chapter 9's optional box showed -1 is a quadratic residue exactly when $p \equiv 1 \pmod 4$, so some x has $p \mid x^2 + 1 = (x + i)(x - i)$. If p stayed prime in $\mathbb{Z}[i]$ it would have to divide one of the factors — but $p \nmid (x \pm i)$, since $\frac{x}{p} \pm \frac{1}{p}i$ is not a Gaussian integer. So p is *not* a Gaussian prime; it factors, and the only way a rational prime factors is $p = (a + bi)(a - bi) = a^2 + b^2$. That is the bridge from "-1 is a square mod p" to "p is a sum of two squares."

What to try when you're stuck

When a problem involves rotations, regularly spaced points, powers of complex numbers, trig identities, or sieving sums by residues:

1. **Switch to the right form.** Use rectangular $a + bi$ for adding, polar $r(\cos\theta + i\sin\theta)$ for multiplying and powers.
2. **Reframe symmetric points as roots of unity.** Regularly spaced points on a circle solve $z^n = 1$; their symmetric sums often collapse via "multiply by ζ to rotate."
3. **Factor $z^n - 1$ by its roots** to turn products of chords or sines into the integer n.
4. **Use de Moivre** to turn $\cos(n\theta)$, $\sin(n\theta)$, and powers into binomial expansions of $(\cos\theta + i\sin\theta)^n$.
5. **Filter with roots of unity** to extract every dth term of a sum: evaluate at the dth roots of unity and average.
6. **To count by residue,** pack the objects into a generating polynomial $P(x)$ (each object as x^{size}) and average P over the dth roots of unity.
7. **For sums of two squares,** read $a^2 + b^2$ as the norm $|a + bi|^2$; norms multiply, so products of such sums stay such sums.

8. **Take real or imaginary parts** at the end — a real answer often hides inside a complex computation.

THE MOVE

One-page recap. Complex numbers fuse geometry and algebra: arrows you add like vectors and multiply by "multiply lengths, add angles." Toolbox: **roots of unity** ($z^n = 1$ gives a regular n-gon summing to 0), **factor** $z^n - 1$ (chord/sine products collapse to n; the pentagon hides φ), **de Moivre** ($(\cos\theta + i\sin\theta)^n = \cos n\theta + i\sin n\theta$, trig identities for free), and **the roots-of-unity filter** (sieve every dth term by averaging over dth roots — and **count objects by residue** by averaging their generating polynomial the same way). A second box, the **Gaussian integers**, reads $a^2 + b^2$ as a norm $|a + bi|^2$, so sums of two squares multiply and the primes $\equiv 1 \pmod 4$ are exactly the ones that split. Take real/imaginary parts to land a real answer. When you see rotation or regular spacing, go complex.

THE NAME FOR IT

Complex numbers and roots of unity

Numbers of the form $a + bi$ with $i^2 = -1$ are the *complex numbers*; the solutions of $z^n = 1$ are the *nth roots of unity*. The power rule $(\cos\theta + i\sin\theta)^n = \cos n\theta + i\sin n\theta$ is *de Moivre's theorem*, and the sieving trick is the *roots-of-unity filter*.

WHERE IT LEADS The complex numbers are the algebraically closed field where *every* polynomial factors (the *Fundamental Theorem of Algebra*), and roots of unity are the simplest characters of a group — the seed of *Fourier analysis* and *representation theory*. The roots-of-unity filter is a finite *discrete Fourier transform*, and the golden ratio in the pentagon is an early thread of *algebraic number theory*.

CURIOUS CARL • EUREKA

Carl slides his protractor into a drawer. "Rotations. This whole time. Just... multiplication. I would like the afternoon back that I spent with this protractor."

SAME BEAST, NEW COSTUME

The same "rotation is multiplication" beast, in disguise across the book:

de Moivre → Roots of unity → The sine product → Roots-of-unity filter → Napoleon's rotations.

Each turns a question about angles, spacing, or symmetry into plain algebra by letting multiplication carry the rotation — and the answer, however complex the path, comes back real.

The practice ladder

THE PRACTICE LADDER *climb at your own pace*

LAUNCH *Warm the engine. Nudges below.*

1. Compute $(1+i)^2$ and $(1+i)^4$, and describe each as a length and angle.
2. Find all four solutions of $z^4 = 1$ and plot them as a square on the unit circle.

CORE BUILDER *Master the central moves.*

1. Use de Moivre to derive the triple-angle formula $\cos 3\theta = 4\cos^3\theta - 3\cos\theta$.
2. Prove that the sum of all nth roots of unity is 0 for every $n \geq 2$, using the geometric-series / rotation argument.
3. Compute $\binom{n}{0} + \binom{n}{2} + \binom{n}{4} + \cdots$ (every *other* term) using the filter with $\omega = -1$, and simplify.

STRETCH *Transfer to unfamiliar ground.*

1. Use the roots-of-unity filter with cube roots to find $\binom{n}{0} + \binom{n}{3} + \binom{n}{6} + \cdots$ in closed form, and evaluate it for $n = 9$.
2. Prove that $\prod_{k=1}^{n-1} 2\sin\frac{k\pi}{n} = n$ by factoring $z^n - 1$ and setting $z = 1$. Verify for $n = 4$.
3. Derive $\cos 36^\circ = \frac{1+\sqrt{5}}{4}$ from the fifth roots of unity, and explain how the golden ratio φ appears.

THE FLASH *No clock — each cracks open in one flash of insight, once you find the angle. Nudges at the back.*

1. What is i^{2026}?
2. Evaluate $\cos 0 + \cos\frac{2\pi}{5} + \cos\frac{4\pi}{5} + \cos\frac{6\pi}{5} + \cos\frac{8\pi}{5}$.

PROOF LAB *Write it clean, in full sentences.*

1. Prove de Moivre's theorem by induction on n, using the angle-addition formulas (or the multiplication rule) for the inductive step.

2. Prove the roots-of-unity filter: for a primitive dth root ω, the average $\frac{1}{d}\sum_{j=0}^{d-1} f(\omega^j x)$ extracts the terms of f whose exponent is divisible by d.

3. Show that exactly 24 of the subsets of $\{1, 2, \dots, 6\}$ have sum divisible by 3, by evaluating $P(x) = \prod_{j=1}^{6}(1 + x^j)$ at the cube roots of unity and averaging.

4. Prove the Brahmagupta–Fibonacci identity $(a^2 + b^2)(c^2 + d^2) = (ac - bd)^2 + (ad + bc)^2$ using $|zw| = |z|\,|w|$ for $z = a + bi$, $w = c + di$; then use it to write $65 = 5 \cdot 13$ as a sum of two squares.

5. Prove that a prime $p \equiv 3 \pmod 4$ is never a sum of two squares. (For contrast, every prime $p \equiv 1 \pmod 4$ is, since it factors as $(a + bi)(a - bi)$ in the Gaussian integers.)

REFLECTION *No nudge. Write a few sentences.*

1. In the cosine-sum payoff, the answer was real but the method was complex. Describe how a "detour through the imaginary plane" simplified a real problem, and when you'd reach into that box again.

NUDGES

Launch. (1) $(1+i)^2 = 2i$ (length 2, angle 90°); $(1+i)^4 = -4$ (length 4, angle 180°). (2) $z = 1, i, -1, -i$ — a square. **Core.** (3) Expand $(\cos\theta + i\sin\theta)^3$ and take the real part. (4) $\zeta S = S$ forces $S = 0$ since $\zeta \neq 1$. (5) Filter with $\omega = -1$: $\frac{2^n + 0^n}{2} = 2^{n-1}$ (for $n \geq 1$). **Stretch.** (6) $\frac{2^n + 2\cos(n\pi/3)}{3}$; for $n = 9$, $\frac{512 + 2\cos 3\pi}{3} = \frac{512-2}{3} = 170$. (7) $\prod_{k=1}^{n-1}(z - \zeta^k) = 1 + z + \dots + z^{n-1}$; at $z = 1$ this is n, and $|1 - \zeta^k| = 2\sin\frac{k\pi}{n}$. For $n = 4$: $2\sin 45^\circ \cdot 2\sin 90^\circ \cdot 2\sin 135^\circ = \sqrt{2}\cdot 2 \cdot \sqrt{2} = 4$. (8) $4c^2 - 2c - 1 = 0 \Rightarrow c = \frac{1+\sqrt{5}}{4} = \frac{\varphi}{2}$. **The Flash.** (9) $i^{2026} = i^{2026 \bmod 4} = i^2 = -1$. (10) 0, the real part of the vanishing root sum. **Proof.** (11) The base case $n = 1$ is the definition. Assuming $(\cos\theta + i\sin\theta)^n = \cos n\theta + i\sin n\theta$, multiply both sides by $\cos\theta + i\sin\theta$: the angle-addition formulas collapse the product to $\cos(n+1)\theta + i\sin(n+1)\theta$, an angle-$n\theta$ arrow turned by a further θ, closing the induction (negative n follows by reciprocals). (12) Writing $f(x) = \sum_m c_m x^m$, the average is $\frac{1}{d}\sum_{j=0}^{d-1} f(\omega^j x) = \sum_m c_m x^m \left(\frac{1}{d}\sum_{j=0}^{d-1}\omega^{jm}\right)$; the inner geometric sum equals 1 when $d \mid m$ and 0 otherwise (since $\omega^m \neq 1$ makes it vanish), so only exponents divisible by d survive. (13) The coefficient of x^s in $P(x) = \prod_{j=1}^{6}(1 + x^j)$ counts subsets of sum s, so the number with $3 \mid s$ is $\frac{1}{3}(P(1) + P(\omega) + P(\omega^2))$ for $\omega = e^{2\pi i/3}$. Here $P(1) = 64$, and grouping the six factors by $j \bmod 3$ gives $P(\omega) = P(\omega^2) = [(1+\omega)(1+\omega^2)]^2 \cdot 2^2 = 1 \cdot 4 = 4$; hence the count is $\frac{64+4+4}{3} = 24$. (14) For $z = a + bi$, $w = c + di$ the modulus is multiplicative, so $(a^2 + b^2)(c^2 + d^2) = |z|^2|w|^2 = |zw|^2$; since $zw = (ac - bd) + (ad + bc)i$, this reads $(ac - bd)^2 + (ad + bc)^2$. Taking $z = 1 + 2i$, $w = 2 + 3i$ gives $zw = -4 + 7i$, so $65 = 5 \cdot 13 = 4^2 + 7^2$ (the conjugate pairing gives $8^2 + 1^2$). (15) Modulo 4 every square is 0 or 1, so a sum of two squares is $\equiv 0, 1, 2 \pmod 4$ and never 3; a prime $p \equiv 3 \pmod 4$ therefore cannot be one — equivalently it stays prime in the Gaussian integers, unlike primes $p \equiv 1$, which split as $(a + bi)(a - bi) = a^2 + b^2$.

MATH LAB #3

Number-Theory Magic Show

Every "trick" in this show is a congruence in a cape. Perform each illusion, then pull back the curtain and name the modular arithmetic doing the work.

Chapters 9 and 10 taught the one move behind every number-theory conjuring act: *stop caring about the number, and care only about its remainder.* In this lab you *perform* four tricks and then explain each with a single line of modular arithmetic. A trick you can explain is a trick you can invent — which is the finale.

CURIOUS CARL • FIRST GUESS

Carl: "To check the trick? I'll run it on a few hundred examples until I'm convinced."

The catch: examples build faith, not proof, Carl. One congruence — what does each piece become *mod* the right number — explains *every* case at once, and shows you why the illusion can never fail.

STATION 1 • the "abcabc" vanish

Ask a volunteer for any three-digit number $\overline{abc}$; write it twice to make the six-digit $\overline{abcabc}$ (e.g. 472 → 472472). **Announce that it is divisible by 7, by 11, and by 13 — and be right every time.** Why does this always work?

Pull the curtain: what is $\overline{abcabc}$ as a multiple of $\overline{abc}$?

STATION 2 • the last-digit prophecy

Have someone pick an exponent n; before they compute, **predict the last digit of 7^n.** *Pull the curtain:* the last digit is 7^n mod 10, and the powers of 7 cycle. Find the cycle and the rule (in terms of n mod 4).

STATION 3 • the mind-read of nine

Have a friend take any whole number N, add up its digits to get S, and compute $N - S$ (kept hidden). Ask them to read you *all but one* of its nonzero digits. **Name the hidden digit.**

Pull the curtain: show $N - S \equiv 0 \pmod 9$, so its digit sum is a multiple of 9; the hidden digit is whatever completes the revealed digits to the next multiple of 9.

STATION 4 • the alternating oracle

Build a divisibility test for 11 and use it to decide, at a glance, whether 918 082 is a multiple of 11. *Pull the curtain:* what is 10 mod 11, and what does that do to $\overline{d_k \dots d_1 d_0} = \sum d_i 10^i$?

STATION 5 • Finale — invent your own trick

Design a brand-new divisibility rule — for 4, or 8, or 37, your choice — and justify it with *one* congruence. (Hint for 4: what is 100 mod 4, and why does that mean only the last two digits matter?)

NUDGES

Peeks (try first!). **(1)** $\overline{abcabc} = \overline{abc} \times 1001$, and $1001 = 7 \times 11 \times 13$ — so the six-digit number is divisible by all three. **(2)** Powers of 7 end in $7, 9, 3, 1, 7, 9, 3, 1, \dots$ (period 4): last digit is $7, 9, 3, 1$ according as $n \equiv 1, 2, 3, 0 \pmod 4$. **(3)** $N \equiv S \pmod 9$ (since $10 \equiv 1$), so $N - S \equiv 0 \pmod 9$; its digits sum to a multiple of 9, and the hidden digit is 9 minus (revealed sum mod 9), or 9 if that is 0. **(4)** $10 \equiv -1 \pmod{11}$, so $N \equiv d_0 - d_1 + d_2 - \cdots$: the *alternating* digit sum. For 918082: $2 - 8 + 0 - 8 + 1 - 9 = -22$, divisible by 11 — yes, it's a multiple of 11. **(5)** For 4: $100 \equiv 0 \pmod 4$, so every digit above the tens place contributes a multiple of 4; only the last two digits decide divisibility by 4.

CURIOUS CARL • EUREKA

Carl: "Oh — there's no magic! Every trick is just 'what does this become mod the right number,' and the remainder does the whole show."

CHAPTER FOURTEEN

14

The Picture Behind the Formula

A square root is a distance, a sum of them is a path, a difference is a triangle inequality, and a quotient is a slope. Draw the algebra — and the minimum you were grinding for becomes a single straight line.

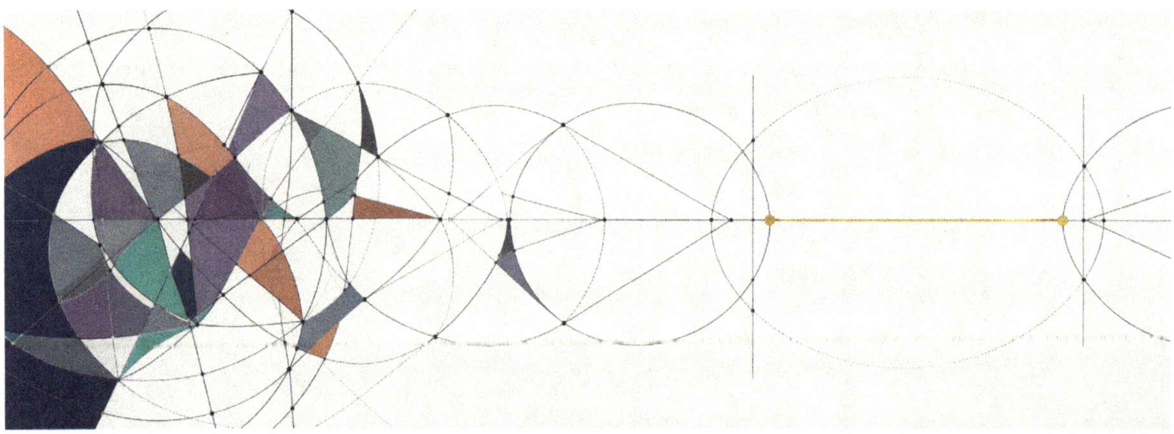

Geometry: draw the algebra.

A pool player solves the next problem by pure instinct — and it hides an ugly algebra optimization inside.

WONDER THIS

You're at point A, your friend is at point B, both on the same side of a long straight wall. You walk to the wall, touch it anywhere, then walk to your friend — using the *shortest total distance.*

Where should you touch the wall? Straight toward the wall and then to B? Aim for the midpoint? There's one perfect spot.

CURIOUS CARL • FIRST GUESS

Carl: "Coordinates, obviously — put the wall on the x-axis, call the touch point $(t, 0)$, write the total distance as two square roots $\sqrt{\cdots} + \sqrt{\cdots}$, and differentiate."

The catch: that derivative of a *sum of two square roots* is genuinely nasty, Carl — but the moment you realize each root is a *distance*, one mirror image turns the whole bent path into a straight one, and the answer falls out with no calculus at all.

Minimizing "distance to the wall plus distance from the wall to B" with calculus is possible but ugly. The slick solution turns the problem inside out: *reflection*. Put the mirror image of your friend on the *far* side of the wall. Call it B'. For any point P on the wall, the **distance** PB exactly equals PB' (the wall is the perpendicular bisector of B and B').

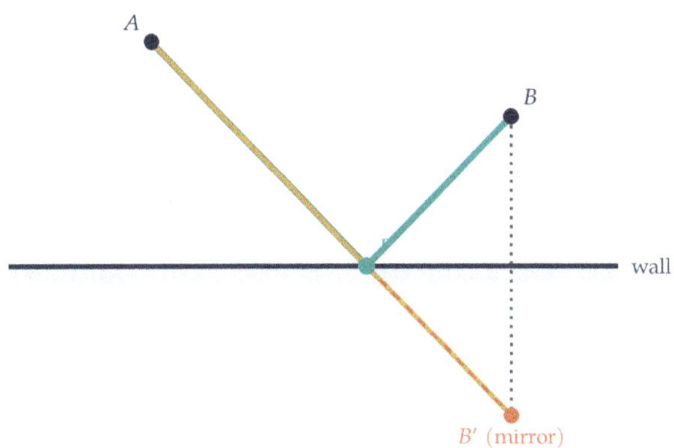

So minimizing $AP + PB$ is the same as minimizing $AP + PB'$ — and the shortest path from A to B' is simply the *straight line*! The best point P is where the segment AB' crosses the wall. **Reflect the destination across the wall, then go straight**. And the law of reflection (angle in = angle out) drops into your lap as a bonus.

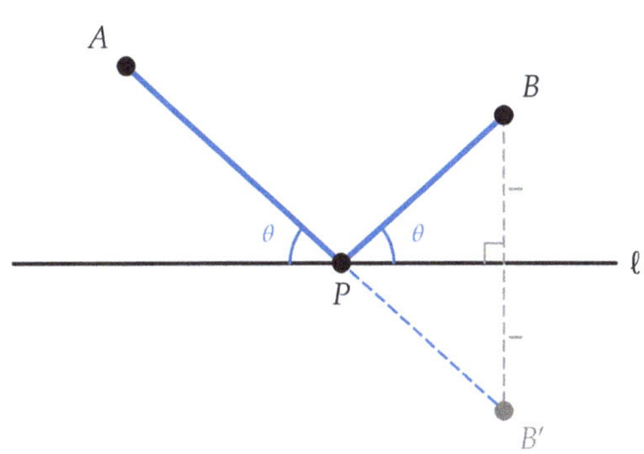

$$AP + PB = AP + PB' = AB' \text{ (shortest)}$$

Reflect B across the line to B': the straight segment AB' meets the line at the optimal P, giving the shortest path $A \to P \to B$.

THE TURN *from grind to insight*

The grind is to plant a coordinate at the touch point, write the total length $\sqrt{\cdots} + \sqrt{\cdots}$ as a function of that one variable, differentiate the sum of two square roots, and solve for where the slope vanishes — a painful derivative for a problem a pool player solves by eye. **The clever picture** is to read each square root as a *distance*: reflect B to B', so the bent path $AP + PB$ becomes the straight path $AP + PB'$, whose minimum is just the segment AB'. It wins because seeing the algebra as *geometry* converts "minimize a sum of radicals" into "a straight line is shortest" — no calculus at all.

THE MOVE

A square root is a distance. The expression $\sqrt{(x-a)^2 + (y-b)^2}$ is exactly the distance from (x, y) to the fixed point (a, b). So an algebraic expression built from such roots is, underneath, a statement about a *picture*:

- a **sum** of two roots is a two-leg *path* — minimize it with a straight line (reflect if the moving point is confined to a line);
- a **difference** of two roots is $|PA - PB|$ — maximize it with the triangle inequality;
- a **quotient** $\frac{y-b}{x-a}$ is a *slope*, and a root under a linear constraint is a *distance to a line*.

Draw the points, and the optimum you were grinding for becomes obvious.

A sum of two radicals is a path

Watch the reflection idea eat a problem that looks like it needs heavy calculus.

WATCH THIS **#1 Minimize a sum of two square roots**

Problem. Over all real x, minimize

$$f(x) = \sqrt{x^2 + 4} + \sqrt{(6-x)^2 + 1}.$$

Read it as a path. Put a moving point $P = (x, 0)$ on the x-axis. Then $\sqrt{x^2+4} = \sqrt{(x-0)^2 + (0-2)^2}$ is the distance PA to $A = (0, 2)$, and $\sqrt{(6-x)^2+1}$ is the distance PB to $B = (6, 1)$. So $f(x) = PA + PB$: a two-leg path from A down to the axis and back up to B.

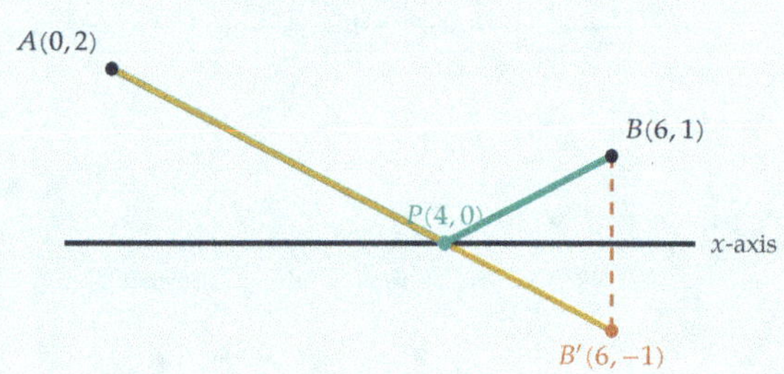

Reflect and go straight. Both points sit above the axis, so reflect B to $B' = (6, -1)$; then $PB = PB'$

and $f(x) = PA + PB'$. The shortest such path is the straight segment AB':

$$f_{\min} = AB' = \sqrt{(6-0)^2 + (-1-2)^2} = \sqrt{36+9} = \sqrt{45} = 3\sqrt{5},$$

achieved where AB' crosses the axis — the line $y = 2 - \frac{x}{2}$ hits $y = 0$ at $x = 4$. **No derivative, no squaring: just a straight line.** (Check: $f(4) = \sqrt{20} + \sqrt{5} = 2\sqrt{5} + \sqrt{5} = 3\sqrt{5}$.) ■

Trigger A sum of square roots $\sqrt{(x-a)^2 + p^2} + \sqrt{(x-c)^2 + q^2}$ to minimize — each root is a distance from a point on a line.

Do first Plant the free point on its line, name the two fixed points, and reflect one across the line; the minimum is the straight-line distance between the reflected pair.

Trapdoor Reflecting the *wrong* way: if the two fixed points are already on *opposite* sides of the line, don't reflect — the straight segment already crosses, and the minimum is just AB.

Upgrade The same picture handles light rays, billiards, and shortest pipelines; the vector version (next) proves the bound $\sqrt{c^2 + (p+q)^2}$ in one line.

CURIOUS CARL • TRAPDOOR!

Carl: "I'll minimize $\sqrt{x^2+4} + \sqrt{(6-x)^2+1}$ by minimizing each root *separately* — the first is smallest at $x = 0$, the second at $x = 6$. So the answer is one of those?"

The catch: the two legs pull against each other — draw one end in and the other stretches, so no single x can minimize both. Greedily optimizing each piece is a trap; it's the *whole* path that has to be short. Reflect one endpoint across the line, and a single straight segment beats every greedy compromise at once.

When the parts add up: the Minkowski move

Sometimes the two radicals aren't "a point on a line" at all — they're two *vectors* whose pieces happen to add to a constant. Then glue them tip-to-tail.

WATCH THIS **#2 Minimize by adding vectors**

Problem. Minimize $g(x) = \sqrt{x^2+9} + \sqrt{(10-x)^2+16}$ over real x.

Read each root as a vector length. Let $\vec{u} = (x, 3)$ and $\vec{v} = (10-x, 4)$, so $\sqrt{x^2+9} = |\vec{u}|$ and $\sqrt{(10-x)^2+16} = |\vec{v}|$. The horizontal parts add to a *constant*: $\vec{u} + \vec{v} = (10, 7)$, no matter what x is. The triangle inequality for vectors — $|\vec{u}| + |\vec{v}| \ge |\vec{u} + \vec{v}|$ — says a two-leg journey is at least as long as the direct hop:

$$g(x) = |\vec{u}| + |\vec{v}| \ \ge\ |\vec{u} + \vec{v}| = |(10,7)| = \sqrt{100+49} = \sqrt{149}.$$

When is it tight? Equality holds when $\vec{u}$ and $\vec{v}$ point the *same* way, i.e. $\frac{x}{3} = \frac{10-x}{4}$, giving $x = \frac{30}{7}$. So $g_{\min} = \boxed{\sqrt{149}}$. ■

This is *Minkowski's inequality*, $\sqrt{a^2+b^2}+\sqrt{c^2+d^2} \geq \sqrt{(a+c)^2+(b+d)^2}$, and it is nothing but the triangle inequality wearing algebra's clothes. Whenever a sum of two square roots has cross-terms that cancel into a fixed total, this is your one-line bound.

A difference is a triangle inequality

Flip the sign and the geometry flips too. A *difference* of distances $PA - PB$ can be made large, and the triangle inequality caps exactly how large.

WATCH THIS #3 Maximize a difference of square roots

Problem. Over all real x, maximize $h(x) = \sqrt{x^2+4} - \sqrt{(x-3)^2+1}$.

Read it as $PA - PB$. With $P = (x, 0)$, $A = (0, 2)$, $B = (3, 1)$, we have $h(x) = PA - PB$. The *reverse* triangle inequality says $PA - PB \leq AB$, with equality exactly when P, B, A are collinear and B lies between P and A — that is, when P sits on line AB, out past B.

$$h(x) \leq AB = \sqrt{(3-0)^2+(1-2)^2} = \sqrt{10}.$$

Line AB is $y = 2 - \frac{x}{3}$; it crosses the axis at $x = 6$, and there $P = (6, 0)$ lies beyond B, so the bound is achieved: $h_{\max} = \boxed{\sqrt{10}}$ at $x = 6$. ■

TRY TO BREAK IT

The reflection trick assumed A and B were on the *same* side of the line. Put them on *opposite* sides instead and try to minimize $AP + PB$. Should you still reflect? Draw it: the straight segment AB already pierces the line, so the minimum is just AB — and reflecting now gives a *longer*, wrong answer. *Where exactly does the "reflect then go straight" argument break when the points start on opposite sides?*

Quotients are slopes; constraints are distances

Two more disguises round out the toolkit. A *quotient* of coordinate differences is a slope, and a root minimized under a *linear constraint* is a distance to a line.

WATCH THIS #4 A constrained minimum is a perpendicular

Problem. Minimize $\sqrt{x^2+y^2}$ subject to $3x + 4y = 25$.

Read it as a distance. $\sqrt{x^2+y^2}$ is the distance from the origin to the point (x, y), and the constraint pins (x, y) to a *line*. The smallest distance from a point to a line is the *perpendicular* distance:

$$\min \sqrt{x^2+y^2} = \frac{|3\cdot 0 + 4\cdot 0 - 25|}{\sqrt{3^2+4^2}} = \frac{25}{5} = 5,$$

reached at the foot of the perpendicular, $(3,4)$. No Lagrange multipliers, no substitution — just "closest point on a line is straight across." ■

The slope cousin is just as handy: to optimize $\frac{y-b}{x-a}$ as (x,y) runs over a curve, read it as the *slope* of the line from the fixed point (a,b) to a moving point on the curve: the extreme values are the slopes of the two *tangent* lines from (a,b). A fraction becomes a picture of a pivoting line.

The bridge runs both ways: power of a point

We have been turning algebra into geometry. The traffic runs the other direction too. Here is a purely geometric fact about circles that turns out to *be* a single algebraic quantity, $d^2 - r^2$. Draw any line through a fixed point P that crosses a circle at two spots; the product of the two distances from P is *always the same*, whichever line you draw.

WATCH THIS #5 Intersecting chords

Claim. Two chords of a circle cross at an interior point P. If one is split into pieces PA and PB, the other into PC and PD, then $PA \cdot PB = PC \cdot PD$.

The auxiliary move: draw two more chords. Connect A to C and B to D, forming triangles $\triangle PAC$ and $\triangle PDB$. They are *similar*: the angles at P are equal (vertical angles), and $\angle CAP = \angle BDP$ as inscribed angles subtending the *same arc* BC.

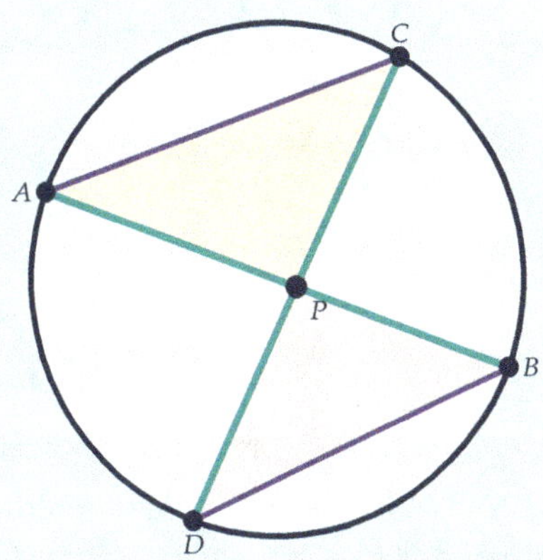

The consequence. Similar triangles have proportional sides: $\frac{PA}{PD} = \frac{PC}{PB}$, so

$$PA \cdot PB = PC \cdot PD. \blacksquare$$

This constant product has a name and a single formula. For a point P at distance d from the center of a circle of radius r, the *signed power* is $\text{Pow}(P) = d^2 - r^2$. With *directed* lengths along a secant, the product of the two directed intersection distances equals $d^2 - r^2$ exactly; with ordinary positive lengths, an *exterior* point gives $PA \cdot PB = d^2 - r^2$ while an *interior* point gives $PA \cdot PB = r^2 - d^2$. Either way, a circle's "arithmetic" at P is a single algebraic number: geometry read back as algebra.

△ **TRAPDOOR**

The power of a point is a *signed* quantity, $\mathrm{Pow}(P) = d^2 - r^2$. With *directed* lengths, the secant product is exactly $d^2 - r^2$. With ordinary *positive* segment lengths: an *exterior* point gives $PA \cdot PB = d^2 - r^2 > 0$ (which also equals the squared tangent length PT^2), while an *interior* point gives $PA \cdot PB = r^2 - d^2 = -(d^2 - r^2)$. Mixing up the sign — or forgetting that an interior point flips it — is the usual stumble; check whether your point is inside or outside first.

WATCH THIS #6 Tangent and secant, unified

Claim. From an external point P: $PT^2 = PA \cdot PB$ for any tangent PT and any secant through A, B. Equivalently, both equal the power $d^2 - r^2$.

Tangent–secant via similar triangles. Connect T to A and T to B. The tangent–chord angle $\angle PTA$ equals the inscribed angle $\angle TBA$ (both cut off arc TA); with the shared angle at P, triangles $\triangle PTA$ and $\triangle PBT$ are similar, so $\frac{PT}{PB} = \frac{PA}{PT}$, i.e. $PT^2 = PA \cdot PB$.
Why it's $d^2 - r^2$. Take the secant through the center: it meets the circle at distances $d - r$ and $d + r$ from P, whose product is $(d - r)(d + r) = d^2 - r^2$. Every secant gives the same product, and the tangent is the limiting secant where A, B merge at T, so $PT^2 = d^2 - r^2$.

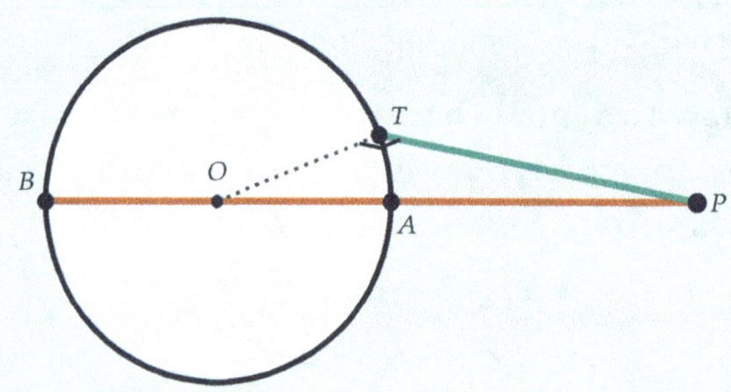

$$PT^2 = PA \cdot PB = d^2 - r^2$$

■

Intersecting chords (interior point), tangent-squared-equals-secant-product (exterior point), and the formula $d^2 - r^2$ are not three theorems but *three views of one quantity*. (One sign caution: for an *interior* point the ordinary chord product reads $|d^2 - r^2|$, while *directed* lengths read the signed $d^2 - r^2$.) *A geometric fact and an algebraic number, secretly the same.*

Squash the ellipse into a circle

An ellipse looks harder than a circle, but it is only a circle in a funhouse mirror. The ellipse $\frac{x^2}{a^2} + \frac{y^2}{b^2} = 1$ is exactly what you get by taking the unit circle $u^2 + v^2 = 1$ and stretching it: $x = au$ across, $y = bv$ up. Run that backward, $u = x/a$ and $v = y/b$, and the ellipse *becomes* a circle. Solve the easy circle problem, then stretch the answer back.

THE MOVE

The squash-and-stretch. The scaling $(x,y) \mapsto (u,v) = (x/a,\ y/b)$ turns $\frac{x^2}{a^2} + \frac{y^2}{b^2} = 1$ into the unit circle $u^2 + v^2 = 1$. Being a *linear* map, it preserves everything "affine":

- straight lines, and *tangency* (a tangent line stays tangent);
- which points lie on / inside / outside the curve;
- *midpoints*, ratios of division along a line, and *parallelism*;
- *ratios of areas* (every area scales by the same factor ab).

So any ellipse question about tangents, chords, midpoints, incidence, or areas becomes the same question on a circle. **Squash to the circle, solve, stretch back.**

△ TRAPDOOR

The squash is not a rigid motion — it *distorts* lengths and angles. It preserves *ratios* of areas but not areas themselves, and it destroys right angles, distances, foci, and, crucially, *arc length*. Never use it on a problem about an ellipse's *perimeter*, its *focal distances*, or a *perpendicularity* condition; those are metric, not affine.

WATCH THIS — #7 The tangent to an ellipse, with no calculus

Problem. Find the tangent line to $\frac{x^2}{a^2} + \frac{y^2}{b^2} = 1$ at a point (x_0, y_0) on it.

Squash. Under $u = x/a$, $v = y/b$, the ellipse becomes $u^2 + v^2 = 1$ and the point becomes $(u_0, v_0) = (x_0/a,\ y_0/b)$. On a circle, the tangent at (u_0, v_0) is the familiar $uu_0 + vv_0 = 1$ (perpendicular to the radius — no derivatives needed).

Stretch back. Substitute $u = x/a,\ v = y/b,\ u_0 = x_0/a,\ v_0 = y_0/b$:

$$\frac{x}{a}\cdot\frac{x_0}{a} + \frac{y}{b}\cdot\frac{y_0}{b} = 1 \quad \Longrightarrow \quad \boxed{\frac{xx_0}{a^2} + \frac{yy_0}{b^2} = 1}.$$

The standard ellipse-tangent formula, which usually costs an implicit derivative, follows straight from the circle's $uu_0 + vv_0 = 1$. ■

WATCH THIS — #8 The largest triangle in an ellipse

Problem. What is the maximum area of a triangle inscribed in $\frac{x^2}{a^2} + \frac{y^2}{b^2} = 1$?

Squash to the circle. Triangles inscribed in the ellipse correspond exactly to triangles inscribed in the unit circle, and *area ratios* are preserved, so the biggest one on the ellipse is the squash of the biggest one on the circle. The largest triangle inscribed in a unit circle is equilateral, with area $\frac{3\sqrt{3}}{4}$.

Stretch back. Every area scales by ab under $(u, v) \mapsto (au, bv)$, so the largest inscribed triangle of the ellipse has area

$$\frac{3\sqrt{3}}{4}\,ab.$$

(For $\frac{x^2}{9} + \frac{y^2}{4} = 1$: $\frac{3\sqrt{3}}{4} \cdot 6 = \frac{9\sqrt{3}}{2}$.) The maximizer on the ellipse is the *stretch* of the equilateral triangle — no longer equilateral, but still maximal, because the squash cannot change which triangle is biggest. ■

The same move settles a whole family of ellipse facts by borrowing the circle's. The *midpoints of a family of parallel chords* lie on a line through the center (the circle's perpendicular-diameter fact, stretched; these are *conjugate diameters*). The *chord of contact* of the two tangents from an external point (x_1, y_1) is $\frac{xx_1}{a^2} + \frac{yy_1}{b^2} = 1$ (the circle's polar line, stretched). And any *ratio* of areas cut by chords transfers untouched. *One squash, and the ellipse gives up its secrets in the circle's own words.* (If the ellipse is *tilted* — an xy cross-term — first shear it upright, then squash; the same affine logic applies, as long as the question never mentions perimeter or angles.)

One squash flattens a triangle, too

The ellipse squash has a twin for straight-sided figures. Because an affine map preserves *ratios* of lengths and *ratios* of areas, it flattens any lopsided triangle into a convenient right one (a vertex at the origin, integer legs) where a cevian-ratio question, or "what fraction is the little central triangle that three cevians cut out?", becomes three line equations and one shoelace. A scalene triangle whose sides are split 1:2, 2:3, and 3:5 hides a central triangle of area exactly $\frac{121}{5814}$ of the whole, and the squash hands it over in about a minute, where balancing all those ratios by hand would take five. **One squash, and a hard ratio problem becomes arithmetic.**

⚠ TRAPDOOR

This is the most seductive move in the chapter, and the one that most needs a fence. An affine map preserves *only* what is built from ratios — division ratios, parallelism, midpoints, collinearity, concurrency, and *ratios* of areas — and quietly destroys everything metric: lengths, angles, perpendicularity, and circles (a circle flattens into an ellipse). So squash a figure to read a *ratio* or an *area-fraction*, and *never* to read an actual length, an angle, a right angle, or anything about a circle. If the problem says "length," "perimeter," "perpendicular," or "circle," stay in the figure you were given. Reading a ratio off the shadow is fair play; reading a length off it is the trap.

SEEN BEFORE?

This squash is the coordinate half of the *mass point* method: mass points *balance* a ratio on a see-saw, the affine shadow *computes* it on a grid. Both — with the area method and Ceva's theorem — fill their own chapter, **Chapter 15** ("The Lever and the Shadow"), where the one-minute central-triangle trick and the parallelogram-to-square flatten are worked in full. The picture-drawing chapter and the ratio-computing chapter are running the very same map.

What to try when you're stuck

When a min/max problem is a mess of square roots, fractions, or constraints:

1. **Read every $\sqrt{(x-a)^2+(y-b)^2}$ as a distance** from (x,y) to (a,b), and plot the fixed points.
2. **A sum of two roots is a path** — minimize with a straight line; **reflect** the moving point's partner if it's confined to a line.
3. **If the pieces add to a constant vector,** use Minkowski (the vector triangle inequality) for the bound.
4. **A difference of roots is $|PA-PB|$** — cap it by AB (reverse triangle inequality).
5. **A quotient $\frac{y-b}{x-a}$ is a slope; a root under a linear constraint is a distance to a line** (drop a perpendicular).
6. **Products of segments from a point to a circle** → power of a point d^2-r^2 (mind the sign).

The translation dictionary. Every move in this chapter is one line of a phrasebook — algebra on the left, the picture on the right:

Algebraic form	Geometric reading
$\sqrt{(x-a)^2+(y-b)^2}$	distance from (x,y) to (a,b)
$\sqrt{x^2+p^2}+\sqrt{(q-x)^2+r^2}$	a broken path — reflect to straighten
$\|PA-PB\|$	bounded by AB (reverse triangle inequality)
$ax+by=c$	a line — drop a perpendicular for distance
d^2-r^2	the signed power of a point
$\frac{x^2}{a^2}+\frac{y^2}{b^2}=1$	an affine image of a circle

Don't reach for calculus on the radicals — draw the path.

THE MOVE

One-page recap. Turn algebra into a picture and the optimum is a straight line. Toolbox: distance ($\sqrt{(x-a)^2+(y-b)^2}$ is how far (x,y) is from (a,b)), reflection (sum of two roots, moving point on a line → shortest path), Minkowski ($\sqrt{a^2+b^2}+\sqrt{c^2+d^2}\geq\sqrt{(a+c)^2+(b+d)^2}$, vectors tip-to-tail), reverse triangle inequality (difference of roots $\leq AB$), distance to a line (constrained root → perpendicular), power of a point (segment products $=d^2-r^2$, signed), and the affine squash (flatten an ellipse into a circle to borrow its tangents and area-ratios — the same map flattens a lopsided triangle for cevian ratios in Chapter 15). When a formula looks unbeatable, ask what picture it is describing.

THE NAME FOR IT

Distance geometry

An expression $\sqrt{(x-a)^2+(y-b)^2}$ is a *Euclidean distance*; the bound $\sqrt{a^2+b^2}+\sqrt{c^2+d^2} \geq \sqrt{(a+c)^2+(b+d)^2}$ is *Minkowski's inequality* (the triangle inequality); $|PA-PB| \leq AB$ is the *reverse triangle inequality*; and the constant segment product d^2-r^2 from a point to a circle is the *power of a point*.

WHERE IT LEADS Reading algebra as distance is the seed of *metric geometry* and *normed vector spaces*: Minkowski's inequality is the triangle inequality for the Euclidean norm, and the same viewpoint turns Cauchy–Schwarz into a statement about angles between vectors. Push further and it becomes *functional analysis*, where "distance" and "perpendicular" organize infinite-dimensional spaces; the circle's d^2-r^2 is an invariant of *inversive geometry*.

CURIOUS CARL • EUREKA

Carl watches a page of snarling square roots melt into one dotted straight line between two dots. "All that armor and you were just *distances* in costume. I plotted a few points and you surrendered — the algebra didn't even swing."

SAME BEAST, NEW COSTUME

The same "read the algebra as a distance" beast, in disguise:

Sum of two roots → Shortest path (reflect) → Minkowski / add vectors → Difference → reverse triangle inequality → Constraint → distance to a line → Circle product $= d^2-r^2$.

Each is a formula that becomes obvious the instant you draw the points it secretly describes — and the answer is a straight line, a perpendicular, or a single number.

SEEN BEFORE?

Minkowski's inequality is the geometric twin of **Chapter 11**'s Cauchy–Schwarz and AM–GM — all three are the algebra of "balance and distance." Reading a point as (x, y), or a distance as $|z-w|$, is **Chapter 13**'s complex-number viewpoint; and "a straight line is shortest" is the extremal principle of **Chapter 3** in metric clothing.

The practice ladder

THE PRACTICE LADDER *climb at your own pace*

LAUNCH *Warm the engine. Nudges below.*

1. Minimize $\sqrt{x^2+1}+\sqrt{(4-x)^2+4}$ over real x by reading it as a shortest path.

2. A circle has radius 5, and a point P lies 13 units from its center. How long is the tangent segment from P to the circle?

CORE BUILDER *Master the central moves.*

1. Minimize $\sqrt{(x-1)^2+4}+\sqrt{(x-5)^2+9}$ by reflecting across the x-axis.

2. Two chords of a circle cross; one is divided into lengths 3 and 8, the other into 4 and x. Find x.

3. Minimize $\sqrt{x^2+y^2}$ subject to $3x+4y=25$, by reading it as a distance from the origin to a line.

STRETCH *Transfer to unfamiliar ground.*

1. Maximize $\sqrt{(x+1)^2+9}-\sqrt{(x-3)^2+1}$, using the reverse triangle inequality.

2. Minimize $\sqrt{x^2+4}+\sqrt{(6-x)^2+9}$ with Minkowski's inequality (glue the vectors $(x,2)$ and $(6-x,3)$ tip-to-tail), and say where equality holds.

3. From an external point, a tangent has length 6 and a secant hits the circle at distances 4 and x. Find x.

4. Find the tangent line to the ellipse $\frac{x^2}{16}+\frac{y^2}{9}=1$ at a point (x_0,y_0) on it, by squashing to the unit circle.

5. A triangle is inscribed in the ellipse $\frac{x^2}{25}+\frac{y^2}{9}=1$. Find its maximum possible area.

THE FLASH *No clock — each cracks open in one flash of insight, once you find the angle. Nudges at the back.*

1. Minimize $\sqrt{x^2+4}+\sqrt{(3-x)^2+4}$.

2. From a point P outside a circle, a secant meets the circle at A then B with $PA=4$ and $AB=5$; a tangent from P touches at T. Find PT.

PROOF LAB *Write it clean, in full sentences.*

1. Prove the reflection principle: for points A,B on the same side of a line ℓ, the point $P\in\ell$ minimizing $AP+PB$ is where segment AB' (with B' the reflection of B across ℓ) meets ℓ. Show $AP+PB\geq AB'$ for all $P\in\ell$, with equality there.

2. Prove Minkowski's inequality $\sqrt{a^2+b^2}+\sqrt{c^2+d^2}\geq\sqrt{(a+c)^2+(b+d)^2}$, and state exactly when equality holds.

3. Prove that the power of a point P equals d^2-r^2 for every line through P, and explain how the tangent case arises as a limit of the secant case.

REFLECTION *No nudge. Write a few sentences.*

1. For one problem above, what *picture* did the algebra turn into (a path? a difference of distances? a distance to a line?), and how did seeing it make the answer obvious?

NUDGES

Launch. (1) $A = (0,1)$, $B = (4,2)$, moving point on the x-axis; reflect $B \to (4,-2)$, so min $= \sqrt{4^2+3^2} = 5$. (2) $\sqrt{13^2-5^2} = \sqrt{144} = 12$. **Core.** (3) $A = (1,2)$, $B = (5,3)$; reflect $B \to (5,-3)$, giving min $= \sqrt{(5-1)^2+(3+2)^2} = \sqrt{41}$. (4) $3 \cdot 8 = 4x$, so $x = 6$. (5) Perpendicular distance from origin to $3x + 4y = 25$ is $\frac{25}{\sqrt{9+16}} = 5$ (foot at $(3,4)$). **Stretch.** (6) $A = (-1,3)$, $B = (3,1)$; max $= AB = \sqrt{16+4} = 2\sqrt{5}$, at $P = (5,0)$ on line AB beyond B. (7) $(x,2) + (6-x,3) = (6,5)$, so min $= \sqrt{36+25} = \sqrt{61}$; equality when $\frac{x}{2} = \frac{6-x}{3}$, i.e. $x = 2.4$. (8) $6^2 = 4x$, so $x = 9$. (9) Squash to the unit circle: the tangent at $(x_0/4, y_0/3)$ is $uu_0 + vv_0 = 1$; stretching back gives $\frac{xx_0}{16} + \frac{yy_0}{9} = 1$. (10) Max inscribed triangle in the unit circle has area $\frac{3\sqrt{3}}{4}$; areas scale by $ab = 5 \cdot 3 = 15$, so the maximum is $\frac{45\sqrt{3}}{4}$. **The Flash.** (11) $A = (0,2)$, $B = (3,2)$; reflect $B \to (3,-2)$, min $= \sqrt{3^2+4^2} = 5$. (12) $PB = PA + AB = 9$, so $PT^2 = PA \cdot PB = 4 \cdot 9 = 36$ and $PT = 6$. **Proof.** (13) Reflect B across ℓ to B'; for every $P \in \ell$ we have $PB = PB'$, so $AP + PB = AP + PB' \geq AB'$ by the triangle inequality. Equality holds exactly when P lies on segment AB' — that is, at $P = AB' \cap \ell$ — which is thus the unique minimizer. (14) Read the terms as vector lengths with $\vec{u} = (a,b)$, $\vec{v} = (c,d)$; the claim is the triangle inequality $|\vec{u}| + |\vec{v}| \geq |\vec{u} + \vec{v}|$, since $\vec{u} + \vec{v} = (a+c, b+d)$. Squaring reduces it to $|\vec{u}|\,|\vec{v}| \geq \vec{u} \cdot \vec{v}$, which is Cauchy–Schwarz, with equality iff $\vec{u}$ and $\vec{v}$ are parallel and point the same way (i.e. $ad = bc$ with matching signs). (15) Take the line through P and center O with $|PO| = d$: it meets the circle at signed distances $d - r$ and $d + r$, whose product is $(d-r)(d+r) = d^2 - r^2$. For any other secant, similar triangles from the inscribed-angle relation force the same product $PA \cdot PB = d^2 - r^2$; letting the two intersection points merge sends the secant to the tangent, where $PA \cdot PB \to PT^2 = d^2 - r^2$.

CHAPTER FIFTEEN

15

The Lever and the Shadow

Stop chasing ratios through a thicket of similar triangles. Hang weights at the corners and let a see-saw find them, slice the figure into little areas that read the ratios straight back — and when both stall, flatten the whole thing into a shadow where coordinates finish the job.

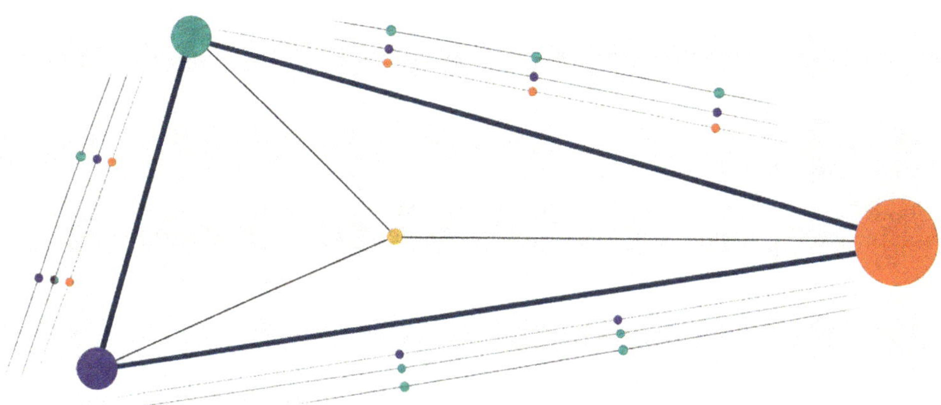

Ratios: hang weights, or cast a shadow.

Here is a ratio question that usually costs a page of similar triangles.

WONDER THIS

In triangle ABC, point D sits on BC with $BD : DC = 2 : 3$, and point E sits on CA with $AE : EC = 1 : 2$. The cevians AD and BE cross at P.

In what ratio does P divide AD? And BE? No similar triangles, no coordinates — is there a way to just *see* it?

CURIOUS CARL • FIRST GUESS

Carl: "Ratios in a triangle? I'll hunt for similar triangles — draw a parallel line through P, find a proportion, chase it around, and...hmm, which line do I even draw?"

The catch: that auxiliary-line hunt is exactly the slog, Carl. Instead, pretend the triangle is a mobile and hang the *right weights* at A, B, C — then every cevian is a balanced see-saw, and the ratios are just the weights read backwards.

The idea is older than algebra: Archimedes' law of the lever. Put a mass m_A at A and a mass m_B at B. They balance at the point P on segment AB where the *torques* match, $m_A \cdot AP = m_B \cdot PB$. So the balance point sits closer to the heavier mass, and

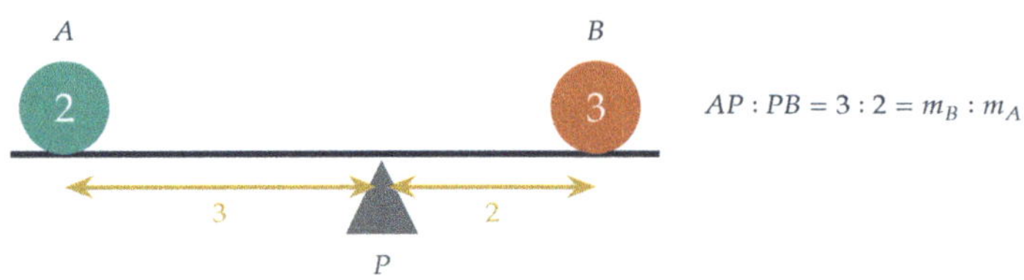

THE MOVE

The mass-point law. Hang a mass at each vertex. Two masses m_A at A and m_B at B balance at the point P on AB with

$$AP : PB = m_B : m_A \qquad \text{(closer to the heavier mass),}$$

and P then carries the combined mass $m_A + m_B$. To solve a cevian problem, *choose the vertex masses so that each given division point is a balance point* — a point D on BC with $BD : DC = p : q$ forces $m_B : m_C = q : p$. Once the masses are set, every cevian is a see-saw and every ratio (including at the crossing point) is read straight off the weights.

That last sentence is the whole method. The art is only in *choosing consistent weights*. Consistency is usually forced on you, one vertex at a time.

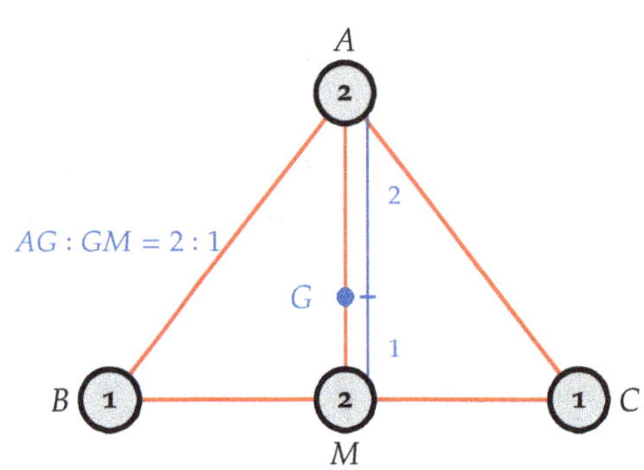

Masses 1 at B and C give the midpoint M mass 2; balanced against mass 2 at A, the centroid sits at $AG : GM = 2 : 1$.

The see-saw law in action

Let's answer the opener with weights.

WATCH THIS #1 Two cevians, read off the weights

Problem. $BD : DC = 2 : 3, AE : EC = 1 : 2$; cevians AD, BE meet at P. Find $AP : PD$ and $BP : PE$.

Balance D **on** BC. We need $m_B : m_C = DC : BD = 3 : 2$. Take $m_B = 3,\ m_C = 2$.

Balance E **on** CA. We need $m_A : m_C = EC : AE = 2 : 1$. Since $m_C = 2$ is already fixed, $m_A = 4$.

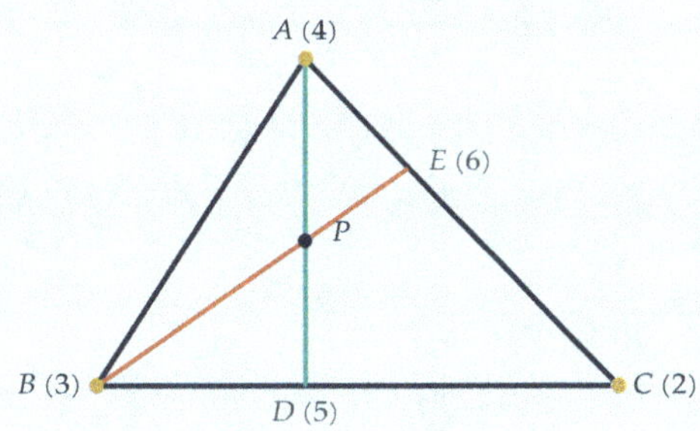

Read the cevians. The balance point D carries $m_D = m_B + m_C = 5$; the balance point E carries $m_E = m_A + m_C = 6$. Now P balances each cevian:

$$AP : PD = m_D : m_A = 5 : 4, \qquad BP : PE = m_E : m_B = 6 : 3 = 2 : 1.$$

Two lines of weighing — no similar triangle drawn. ■

Trigger A ratio along a cevian, or the ratio in which two cevians cut each other, given how the cevians split the sides.

Do first Assign vertex masses inverse to each side-split ($BD : DC = p : q \Rightarrow m_B : m_C = q : p$); the crossing point splits each cevian as (far-end mass) : (vertex mass).

Trapdoor Getting the lever backwards — the ratio is the *inverse* of the masses, since the balance point hugs the heavier side.

Upgrade Scale all masses at once (they're only defined up to a common multiple); pick the smallest integers that make every side-split consistent.

Notice how the second vertex mass was *forced*: once $m_C = 2$ was pinned by the first split, the second split had no freedom left. That is the usual rhythm: set one weight, and the rest cascade. When they cascade consistently, you win; when they *clash*, the diagram is telling you something (as we'll see).

A third weight: Ceva's theorem

Add a third cevian CF and a natural question appears: when do all three cevians pass through *one* point? Mass points answer it in a line.

WATCH THIS #2 Ceva's theorem, by balancing

Claim. In triangle ABC, the cevians AD, BE, CF (with D, E, F on BC, CA, AB) are *concurrent* if and only if

$$\frac{BD}{DC} \cdot \frac{CE}{EA} \cdot \frac{AF}{FB} = 1.$$

Why weights force it. Choose masses so that D and E are balance points: $m_B : m_C = DC : BD$ and $m_C : m_A = EA : CE$. Both cevians AD and BE then pass through the whole system's center of mass

$$G = \frac{m_A A + m_B B + m_C C}{m_A + m_B + m_C},$$

because G lies on the line from each vertex to the balance point of the other two. The *third* cevian CF passes through G exactly when F is the balance point of A and B, i.e. $AF : FB = m_B : m_A$. Multiplying the three side-ratios,

$$\frac{BD}{DC} \cdot \frac{CE}{EA} \cdot \frac{AF}{FB} = \frac{m_C}{m_B} \cdot \frac{m_A}{m_C} \cdot \frac{m_B}{m_A} = 1,$$

so the product equals 1 exactly when F lands on the balance point. This settles *both* directions at once. If the three cevians are concurrent, then F must be that common balance point, forcing the product to 1. Conversely, if the product is 1, then $\frac{AF}{FB} = \frac{m_B}{m_A}$ makes F the balance point of A and B, so CF passes through the same center of mass G as AD and BE — the three cevians concur. ■

The cleanest special case is the one you already know: the three *medians*. Each side is split $1 : 1$, so equal masses $1, 1, 1$ balance everywhere, and the center of mass is the *centroid*. Each median carries a vertex mass 1 against a midpoint mass $1 + 1 = 2$, so

$$(\text{vertex to centroid}) : (\text{centroid to midpoint}) = 2 : 1.$$

The famous "medians meet two-thirds of the way down" is just three equal weights finding their balance. *The centroid is the triangle's center of mass — literally where a cardboard triangle balances on a pin.*

THE TURN *from grind to insight*

The old reflex for a cevian ratio is to manufacture similar triangles: draw an auxiliary parallel, hunt for equal angles, set up a proportion, and chase it across the figure — a different clever line for every problem, and a fresh chance to slip. **The better idea** is to stop drawing lines and start *hanging weights*. Assign masses inverse to the side-splits, and the entire tangle of ratios collapses into arithmetic on the corners: each crossing point is a see-saw, and its ratio is the two masses read backwards. The method wins because it replaces a search (which auxiliary line?) with a recipe (which weights?) — and the recipe never varies.

CURIOUS CARL • TRAPDOOR!

Carl weighs a triangle for a transversal — a line that cuts AB and AC but *misses* vertex B and C entirely. He hangs masses at A, B, C and waits for a balance point that never comes.

The stubborn line speaks up: *"I'm not a cevian — I don't pass through any corner, so I'm nobody's see-saw. Plain weights can't hold me."* Right, Carl: basic mass points balance *cevians*. A transversal needs a heavier tool (a *split mass*, or Menelaus) — or you can sidestep the whole thing with coordinates.

Areas do the weighing: the area method

Mass points read ratios of *lengths*. But the same figures hide ratios of *areas*. And there is a method that reads those directly, so cleanly that it will even hand back the weights we were hanging.

THE MOVE

The area method. Three lemmas turn every ratio into an area, and back:

- *Same height.* If D lies on line BC, then $[ABD] : [ACD] = BD : DC$ — triangles with the same apex A have areas in the ratio of their bases.
- *Shared vertex.* If P lies on AB and Q on AC, then $\dfrac{[APQ]}{[ABC]} = \dfrac{AP}{AB} \cdot \dfrac{AQ}{AC}$ — a corner triangle is the product of its two side-fractions.
- *Cevian to a point.* For a point P inside the triangle with $D = AP \cap BC$, $\dfrac{BD}{DC} = \dfrac{[ABP]}{[ACP]}$ — the side-split equals the ratio of the two areas its cevian cuts off.

The third lemma is the hinge, and it reveals something lovely. For a point P inside ABC, write the three *corner areas*

$$x = [PBC], \qquad y = [PCA], \qquad z = [PAB], \qquad x + y + z = [ABC].$$

The cevian from A through P splits BC as $z : y$; from B, it splits CA as $x : z$; from C, it splits AB as $y : x$. Compare the see-saw: a mass point P with vertex masses m_A, m_B, m_C splits BC as $m_C : m_B$. The two agree exactly when

$$m_A : m_B : m_C = [PBC] : [PCA] : [PAB].$$

The weights you hung at the corners were the areas of the opposite little triangles all along. Mass points and the area method are one idea in two languages (balance, or area) and either can summon the other.

WATCH THIS #3 The inscribed triangle: cut the corners, reveal the center

Problem. On the sides of triangle ABC, take D on BC, E on CA, F on AB, each dividing its side $1 : 2$ in the same rotational sense ($BD : DC = CE : EA = AF : FB = 1 : 2$). What fraction of ABC is the inner triangle DEF?

Cut the three corners. By the shared-vertex lemma, the corner at A (triangle AFE, with F on

AB, E on CA) is

$$\frac{[AFE]}{[ABC]} = \frac{AF}{AB} \cdot \frac{AE}{AC} = \frac{1}{3} \cdot \frac{2}{3} = \frac{2}{9},$$

since $AF : FB = 1 : 2$ gives $\frac{AF}{AB} = \frac{1}{3}$, and $CE : EA = 1 : 2$ gives $\frac{AE}{AC} = \frac{2}{3}$. Rotational symmetry makes the corners at B and C the same, $\frac{2}{9}$ each.

What's left is the center.

$$\frac{[DEF]}{[ABC]} = 1 - 3 \cdot \frac{2}{9} = \frac{1}{3}.$$

Trisect every side, connect the dots, and exactly a *third* of the triangle survives in the middle — no coordinates, just three corner products. ■

WATCH THIS — #4 The one-seventh triangle, by pure areas

Problem. Cevians AD, BE, CF cut the sides $2 : 1$ in the same sense ($BD : DC = CE : EA = AF : FB = 2 : 1$). They miss concurrence and bound a central triangle; show it is exactly $\frac{1}{7}$ of ABC — the answer the affine shadow gave, now with no coordinates.

Weigh the first vertex with areas. Let $P = AD \cap BE$ and apply the cevian-to-a-point lemma on *both* cevians through P:

$$\frac{BD}{DC} = \frac{[ABP]}{[ACP]} = \frac{2}{1}, \qquad \frac{CE}{EA} = \frac{[BCP]}{[ABP]} = \frac{2}{1}.$$

So $[ABP] = 2[ACP]$ and $[BCP] = 2[ABP] = 4[ACP]$; summing to $[ABC]$, the corner areas at P are

$$[PBC] : [PCA] : [PAB] = 4 : 1 : 2 \quad \text{(out of 7)}.$$

Spin the symmetry. The rotation $A \to B \to C$ fixes the whole configuration, so the other central vertices carry the cyclically shifted areas: $Q = BE \cap CF$ is $(2, 4, 1)$ and $R = CF \cap AD$ is $(1, 2, 4)$, each over 7.

Read the center off the corners. Those triples are the *areal coordinates* of P, Q, R, and the area of a triangle in areal coordinates is the determinant of its normalized corner areas — the areal cousin of the shoelace formula:

$$\frac{[PQR]}{[ABC]} = \frac{1}{7^3} \left| \det \begin{pmatrix} 4 & 1 & 2 \\ 2 & 4 & 1 \\ 1 & 2 & 4 \end{pmatrix} \right| = \frac{49}{343} = \frac{1}{7}.$$

Seven regions, one of them the center; the surprising "one-seventh" falls straight out of the corner areas. ■

△ TRAPDOOR

The shared-vertex lemma $\frac{[APQ]}{[ABC]} = \frac{AP}{AB} \cdot \frac{AQ}{AC}$ needs P and Q on the *two sides meeting at the same vertex* A. It is not a free "multiply the fractions" rule: if the two points sit on sides that share no vertex, there is no such product, and you fall back on the same-height lemma or on coordinates. Always check that the little triangle really shares a whole angle with the big one.

SEEN BEFORE?

The areal coordinates $[PBC] : [PCA] : [PAB]$ are the mass-point weights of this chapter's first half and the barycentric coordinates of the formal-term box below — three names for one triple. And the determinant that closed the one-seventh is the areal twin of the shoelace formula from the affine method: both read an area off coordinates, one barycentric, one Cartesian.

Where the see-saw stalls

Balancing and area-chasing are gorgeous when the numbers are kind, but they have a horizon. Basic mass points say nothing about a *transversal* crossing two sides without touching a vertex. The area method needs a clean angle or a symmetry to exploit; strip those away and its bookkeeping balloons. And (the sharpest limit) once the three side-splits are ugly and unequal, so that the cevians are *not* concurrent and carry no symmetry, there is no tidy set of weights and no symmetric area chase to lean on. The instant the Ceva product is not 1 *and* the ratios refuse to line up, the central triangle's corners are just three messy intersections.

WONDER THIS

A triangle carries three cevians with $BD : DC = 1 : 2$, $AE : EC = 2 : 3$, $AF : FB = 3 : 5$. The product $\frac{BD}{DC} \cdot \frac{CE}{EA} \cdot \frac{AF}{FB} = \frac{1}{2} \cdot \frac{3}{2} \cdot \frac{3}{5} = \frac{9}{20} \neq 1$, so the cevians are *not* concurrent — they enclose a central triangle STR.

What fraction of the triangle's area is STR? Mass points can grind out any one corner's ratio, but pinning all three and then the area is a five-minute ordeal. Is there a one-minute route?

There is, and it comes from throwing away everything mass points cling to (the specific triangle) and keeping only what the question actually depends on.

Cast a shadow: the affine escape

A general triangle is inconvenient; its corners land on ugly numbers. But you get to *choose* the triangle, because ratios don't care which one you draw.

THE MOVE

Any triangle is every triangle. An *affine map* $\mathbf{x} \mapsto M\mathbf{x} + \mathbf{b}$ (a linear stretch-and-shear plus a shift, M invertible) carries any triangle onto any other and preserves exactly what a ratio problem needs:

- *collinearity* — lines map to lines, so cevians stay cevians;

- *the ratio a point divides a segment* — every split is unchanged;
- *the ratio of any two areas* — since every area scales by the same factor $|\det M|$.

Lengths and angles are wrecked, but they never appear in a ratio or area-ratio question. So replace the given triangle by the most convenient one of its family — a *right triangle with a vertex at the origin and integer legs* — solve with plain coordinates, and the ratios and area-fractions transfer back untouched.

Why it holds is one line: an affine map respects every barycentric blend, $f(\alpha P + \beta Q) = \alpha f(P) + \beta f(Q)$ when $\alpha + \beta = 1$, so it moves the point dividing PQ in a ratio to the point dividing the images in the *same* ratio; and it multiplies all areas by the constant $|\det M|$, so area-ratios survive. (Affine maps are exactly the projective transformations that fix the line at infinity — so this is the gentlest slice of projective geometry, casting the triangle as a convenient *shadow* of itself.)

WATCH THIS #5 The central triangle, in one minute

Problem. The opener: $BD : DC = 1 : 2$, $AE : EC = 2 : 3$, $AF : FB = 3 : 5$; find $[STR]$ as a fraction of $[ABC]$.

Normalize so the leg points are integers. The split on AB is $3 : 5$ (denominator 8) and on AC is $2 : 3$ (denominator 5), so choose

$$A = (0,0), \qquad B = (8,0), \qquad C = (0,5),$$

making $F = (3,0)$ and $E = (0,2)$ exactly, and $D = (\frac{16}{3}, \frac{5}{3})$ on BC.

Three lines.

$$AD:\ 5x - 16y = 0, \qquad BE:\ x + 4y = 8, \qquad CF:\ 5x + 3y = 15.$$

Three corners, each a 2×2 solve:

$$S = AD \cap BE = \left(\tfrac{32}{9}, \tfrac{10}{9}\right), \quad T = BE \cap CF = \left(\tfrac{36}{17}, \tfrac{25}{17}\right), \quad R = CF \cap AD = \left(\tfrac{48}{19}, \tfrac{15}{19}\right).$$

Shoelace, then rescale. The shoelace formula gives $[STR] = \frac{1210}{2907}$, while $[ABC] = \frac{1}{2} \cdot 8 \cdot 5 = 20$. The ratio is affine-invariant, so

$$\frac{[STR]}{[ABC]} = \frac{1210/2907}{20} = \frac{121}{5814} \approx 0.0208.$$

Three equations, three quick solves, one shoelace — done before the mass-point weights are even hung. ■

The same license flattens a *parallelogram*: every parallelogram is an affine image of the unit square, so a ratio or area question can be solved on the square and read back: no auxiliary triangles, the very things a mass-point solution has to invent for a parallelogram.

WATCH THIS #6 A parallelogram ratio, on the unit square

Problem. In parallelogram $ABCD$, E is the midpoint of side BC and F divides CD with $CF : FD = 1 : 2$. Lines AE and BF meet at P. Find $AP : PE$.

Flatten to the square. Send $ABCD$ to $A = (0,0), B = (1,0), C = (1,1), D = (0,1)$; then $E = (1, \frac{1}{2})$ and $F = (\frac{2}{3}, 1)$.
Two lines. $AE : y = \frac{1}{2}x$. BF : through $(1,0)$ and $(\frac{2}{3}, 1)$, i.e. $y = -3x + 3$.
Cross them. $\frac{1}{2}x = -3x + 3 \Rightarrow x = \frac{6}{7}$, so $P = (\frac{6}{7}, \frac{3}{7})$, which is $\frac{6}{7}$ of the way from A to E:

$$AP : PE = 6 : 1,$$

true in every parallelogram, no auxiliary line drawn. ∎

△ **TRAPDOOR**

Affine normalization is licensed only for what affine maps preserve: division ratios, parallelism, midpoints, collinearity, concurrency, and *ratios* of areas. It scrambles lengths, angles, perpendicularity, and circles (a general circle becomes an ellipse). So never normalize a problem that asks for an actual distance, an angle, a right angle, or anything about a circle. If the wording says "length," "perimeter," "perpendicular," or "circle," stay in the figure you were given — or reach back for mass points and similar triangles.

TRY TO BREAK IT

Mass points and Ceva are two faces of one coin. Take three cevians and *tune* them: fix $BD : DC = 2 : 1$ and $CE : EA = 2 : 1$, then slide F along AB. Watch the Ceva product $\frac{BD}{DC} \cdot \frac{CE}{EA} \cdot \frac{AF}{FB} = 4 \cdot \frac{AF}{FB}$ pass through 1 exactly when $AF : FB = 1 : 4$ — the single instant the three cevians snap into concurrence. Just off that value, they spring apart into a central triangle whose area the affine method can measure but mass points cannot. Feel where "one balance point" becomes "a little triangle," and you've felt the boundary between the two tools.

GOING FURTHER *optional • beyond AIME*

Optional, beyond the contest core. A transversal that misses the vertices — the case that stumped Carl — is governed by *Menelaus's theorem*: if a line crosses BC, CA, AB (extending sides as needed) at X, Y, Z, then

$$\frac{BX}{XC} \cdot \frac{CY}{YA} \cdot \frac{AZ}{ZB} = -1$$

with *signed* ratios (a point outside a segment counts negative). Mass points can still handle it, but only with a *split mass*: a point on a side is given two masses, one for each cevian it must balance, and the extra bookkeeping is exactly why coordinates are often faster here. Menelaus (for transversals) and Ceva (for concurrent cevians) are the two poles of triangle ratio geometry — one line cutting across, three lines meeting at a point — and the affine shadow answers both without a single case of signs.

What to try when you're stuck

The three tools are not rivals but a *decision tree*, and choosing well is the whole creative act:

Which lens should I use?

- *All three cevians meet at a point?* Hang masses. The see-saw finds every ratio.
- *The problem asks for an* ***area*** *(a region, not just a segment)?* Cut into areas. Corner areas *are* the masses, and they subtract cleanly.
- *The diagram is lopsided but only* ***ratios*** *or* ***area-ratios*** *matter?* Flatten it by an affine map to the friendliest triangle, then read the answer off coordinates.

Don't chase similar triangles — hang weights, slice areas, or flatten the figure.

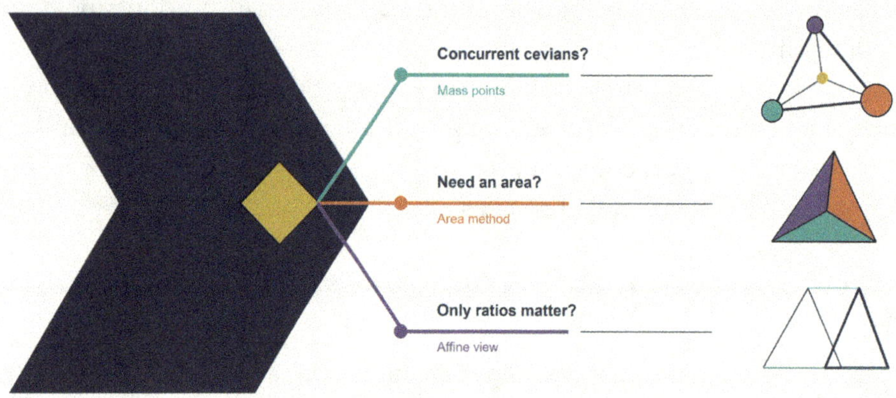

When a triangle or parallelogram problem asks for a ratio of lengths or of areas:

1. **Hang weights first.** If the lines are cevians (through vertices), assign vertex masses inverse to the side-splits and read the ratios off the balance points.
2. **Force the weights to cascade:** set one vertex mass, let the splits fix the rest, and pick the smallest consistent integers.
3. **For concurrency, use Ceva:** the product $\frac{BD}{DC} \cdot \frac{CE}{EA} \cdot \frac{AF}{FB} = 1$ iff the three cevians meet.
4. **For areas, switch to the area method:** $[ABD] : [ACD] = BD : DC$, a corner triangle is $\frac{AP}{AB}\frac{AQ}{AC}$ of the whole, and the corner areas $[PBC] : [PCA] : [PAB]$ are the masses — subtract corners to find an inner region.
5. **When balance and symmetry stall** — a transversal, an area ratio, or non-concurrent cevians — **cast a shadow:** affine-normalize to a right triangle with integer legs (or a parallelogram to the unit square), then use coordinates.
6. **Never normalize** a length, angle, perpendicularity, or circle question — affine maps destroy those.

THE MOVE

One-page recap. Three languages for one thing — the ratios and areas inside a triangle. **Mass points**: hang masses inverse to the side-splits ($BD : DC = p : q \Rightarrow m_B : m_C = q : p$); a balance point carries the summed mass, and a crossing splits each cevian as (far mass):(vertex mass). **Ceva**: three cevians concur iff the product of the three side-ratios is 1 (medians: masses 1, 1, 1, centroid, $2 : 1$). **Area method**: $[ABD] : [ACD] = BD : DC$ (same height), $\frac{[APQ]}{[ABC]} = \frac{AP}{AB}\frac{AQ}{AC}$ (shared vertex), and $\frac{BD}{DC} = \frac{[ABP]}{[ACP]}$ (cevian to a point) — and the corner areas $[PBC] : [PCA] : [PAB]$ *are* the masses; it delivers the one-seventh triangle with no coordinates. **Affine normalization**: when balance and symmetry both stall (ugly unequal splits, general areas), redraw the triangle as a convenient right triangle — ratios and area-ratios survive the squash — and finish with coordinates and the shoelace formula. Balance it, slice it, or cast its shadow.

THE NAME FOR IT

Mass points and affine invariance

Assigning balancing masses to vertices is the *mass point* (or *barycentric-coordinate*) method; the concurrency test $\frac{BD}{DC}\frac{CE}{EA}\frac{AF}{FB} = 1$ is *Ceva's theorem* (its transversal twin, with product -1, is *Menelaus's theorem*). A map $\mathbf{x} \mapsto M\mathbf{x} + \mathbf{b}$ is an *affine transformation*, and the quantities it preserves — division ratios and area ratios — are the *affine invariants.*

WHERE IT LEADS Mass points are the physicist's name for *barycentric coordinates*: writing a point as a weighted average $\frac{m_A A + m_B B + m_C C}{m_A + m_B + m_C}$ is the coordinate system in which Ceva, Menelaus, and the centroid become one-line algebra. The affine viewpoint is *affine geometry* — the study of exactly the properties (ratio, parallelism, area-ratio) that survive $\mathbf{x} \mapsto M\mathbf{x} + \mathbf{b}$ — sitting one rung below *projective geometry*, where even "the line at infinity" becomes movable.

CURIOUS CARL • EUREKA

Carl has stopped chasing ratios. He sets a small weight on each corner of the triangle and watches the whole thing settle.

"So I never really *solve* for the ratio," he says slowly. "I hang weights until the point holds still — and the ratio is just what kept it balanced."

Yes. And on the day the see-saw jams, you flatten the triangle into an easy shadow and read the answer straight off the grid.

SAME BEAST, NEW COSTUME

The same "find the ratio without similar triangles" beast, in disguise:

Lever law → Mass points → Ceva concurrency → Area method → Affine shadow → Barycentric coordinates.

Each refuses to hunt for an auxiliary line. It either balances the ratios on a see-saw or flattens the figure into coordinates — and the proportion falls out as arithmetic.

SEEN BEFORE?

The see-saw law is the *extremal / balance* instinct of **Chapter 11** (the optimum lives where things balance) turned into a tool, and the "center of mass" average is **Chapter 18**'s linearity in geometric clothing. The affine shadow is the same map that squashed an ellipse into a circle in **Chapter 14** — there it borrowed the circle's tangents, here it borrows the coordinate grid's ease.

The practice ladder

THE PRACTICE LADDER *climb at your own pace*

LAUNCH *Warm the engine. Nudges below.*

1. Masses 2 at A and 7 at B hang on segment AB. In what ratio does the balance point P divide AB?
2. A point D on BC has $BD : DC = 3 : 4$. What masses at B and C (smallest integers) make D their balance point?

CORE BUILDER *Master the central moves.*

1. In triangle ABC, D is the midpoint of BC and E lies on AC with $AE : EC = 1 : 2$. Cevians AD and BE meet at P. Find $AP : PD$ and $BP : PE$.
2. Cevians AD, BE, CF have $BD : DC = 2 : 1$ and $CE : EA = 1 : 2$. For what ratio $AF : FB$ are the three cevians concurrent? Use Ceva.
3. Using equal masses $1, 1, 1$, show that the three medians of a triangle meet at a point dividing each median in the ratio $2 : 1$ from the vertex.

STRETCH *Transfer to unfamiliar ground.*

1. In triangle ABC, D lies on BC with $BD : DC = 1 : 2$ and F lies on AB with $AF : FB = 1 : 3$. Cevians AD and CF meet at P. Find $AP : PD$ and $CP : PF$.
2. Points on the three sides of a triangle divide them each in ratio $2 : 1$, taken in the same rotational sense ($BD : DC = CE : EA = AF : FB = 2 : 1$). The three cevians bound a central triangle. Find its area as a fraction of the whole, by affine normalization.
3. In parallelogram $ABCD$, E is the midpoint of BC and F divides CD with $CF : FD = 1 : 2$. Lines AE and BF meet at P. Find $AP : PE$ by flattening to the unit square.

THE FLASH *No clock — each cracks open in one flash of insight, once you find the angle. Nudges at the back.*

1. Masses 5 at A and 3 at B: in what ratio is AB divided at the balance point?

2. In triangle ABC, D on BC has $BD : DC = 3 : 4$ and E on AC has $AE : EC = 3 : 4$. Cevians AD, BE meet at P. Find $BP : PE$.

PROOF LAB *Write it clean, in full sentences.*

1. Prove the see-saw law: if masses m_A at A and m_B at B balance at a point P on segment AB, then $AP : PB = m_B : m_A$. (Use the center of mass, or the torque condition $m_A \cdot AP = m_B \cdot PB$.)

2. Prove Ceva's theorem: cevians AD, BE, CF are concurrent if and only if $\frac{BD}{DC} \cdot \frac{CE}{EA} \cdot \frac{AF}{FB} = 1$, using mass points.

3. Prove that an affine map $\mathbf{x} \mapsto M\mathbf{x} + \mathbf{b}$ preserves (a) the ratio in which a point divides a segment and (b) the ratio of the areas of two triangles — the two facts that make affine normalization valid.

4. *Area method.* In triangle ABC of area 1, D lies on BC with $BD : DC = 1 : 3$ and E lies on CA with $CE : EA = 1 : 1$. Using the same-height and shared-vertex lemmas, find $[ABD]$ and $[CDE]$.

5. *Corner-cutting.* Points D, E, F divide the sides of triangle ABC each in ratio $1 : 2$ in the same rotational sense. Find $[DEF]$ as a fraction of $[ABC]$ by cutting the three corners.

6. *Cevian areas.* In triangle ABC of area 1, cevian AD has $BD : DC = 1 : 2$ and cevian CE (with E on AB) has $AE : EB = 3 : 1$; the two meet at P. Using the cevian-to-a-point lemma, find $[APC]$.

REFLECTION *No nudge. Write a few sentences.*

1. Mass points and coordinates are two languages for the same ratios. Describe one problem where hanging weights felt magical, and one where the see-saw stalled and casting a shadow was the escape. What told you which tool to reach for?

NUDGES

Launch. (1) $AP : PB = m_B : m_A = 7 : 2$. (2) $m_B : m_C = DC : BD = 4 : 3$, so B gets 4, C gets 3. **Core.** (3) D midpoint $\Rightarrow m_B = m_C = 1$; $AE : EC = 1 : 2 \Rightarrow m_A = 2$. Then $AP : PD = m_D : m_A = 2 : 2 = 1 : 1$ and $BP : PE = m_E : m_B = 3 : 1$. (4) Ceva: $\frac{2}{1} \cdot \frac{1}{2} \cdot \frac{AF}{FB} = 1 \Rightarrow AF : FB = 1 : 1$ (F the midpoint). (5) Equal masses 1, 1, 1 give each midpoint mass 2; each median splits as $m_{\text{vertex}} : m_{\text{mid}}$ reversed $= 2 : 1$. **Stretch.** (6) $BD : DC = 1 : 2 \Rightarrow m_B = 2, m_C = 1$; $AF : FB = 1 : 3 \Rightarrow m_A = 6$. Then $AP : PD = m_D : m_A = 3 : 6 = 1 : 2$ and $CP : PF = m_F : m_C = 8 : 1$. (7) Affine-normalize; the central triangle is $\frac{1}{7}$ of the whole (Routh's classic one-seventh). (8) $AP : PE = 6 : 1$ on the unit square. **The Flash.** (9) $AP : PB = 3 : 5$. (10) $m_C = 3, m_B = 4$ (from $3 : 4$), $m_A = 4$ (from $AE : EC = 3 : 4$); $BP : PE = m_E : m_B = 7 : 4$. **Proof.** (11) Place masses m_A at A and m_B at B; their center of mass is $P = \frac{m_A A + m_B B}{m_A + m_B}$. Rewriting $P = A + \frac{m_B}{m_A + m_B}(B - A)$ shows P lies on segment AB with $AP = \frac{m_B}{m_A + m_B} AB$ and $PB = \frac{m_A}{m_A + m_B} AB$, so $AP : PB = m_B : m_A$. Equivalently, the torque balance $m_A \cdot AP = m_B \cdot PB$ yields the same ratio. (12) Assign masses m_A, m_B, m_C so that D balances BC and E balances CA, forcing $\frac{BD}{DC} = \frac{m_C}{m_B}$ and $\frac{CE}{EA} = \frac{m_A}{m_C}$. The center of mass $G = \frac{\sum m_i P_i}{\sum m_i}$ then lies on both AD and BE (combine the two far masses at their foot first), so those cevians meet at G; the third cevian CF passes through G exactly when F balances AB, i.e. $\frac{AF}{FB} = \frac{m_B}{m_A}$.

Multiplying, $\frac{BD}{DC} \cdot \frac{CE}{EA} \cdot \frac{AF}{FB} = \frac{m_C}{m_B} \cdot \frac{m_A}{m_C} \cdot \frac{m_B}{m_A} = 1$, which is Ceva's concurrency condition. (13) Write $\phi(\mathbf{x}) = M\mathbf{x}+\mathbf{b}$. For $P = (1-t)A+tB$, linearity gives $\phi(P) = (1-t)\phi(A)+t\phi(B)$, so ϕ preserves the ratio $t : (1-t)$ in which P divides the segment. A triangle on edge vectors $B-A, C-A$ has area $\frac{1}{2}|\det(B-A,\ C-A)|$; ϕ sends these to $M(B-A), M(C-A)$, scaling every determinant — hence every area — by the same factor $|\det M|$, so any ratio of two areas is unchanged. (14) By the same-height lemma, ABD and ABC share the altitude from A, so $[ABD] = \frac{BD}{BC}[ABC] = \frac{1}{4}\cdot 1 = \frac{1}{4}$. By the shared-vertex lemma at C, $[CDE] = \frac{CD}{CB} \cdot \frac{CE}{CA}[ABC] = \frac{3}{4} \cdot \frac{1}{2} = \frac{3}{8}$. (15) Each cut-off corner triangle spans two sides in ratios $\frac{1}{3}$ and $\frac{2}{3}$, so by the shared-vertex lemma its area is $\frac{1}{3} \cdot \frac{2}{3}[ABC] = \frac{2}{9}$. The three corners have equal area by the rotational symmetry of the construction, so $[DEF] = 1-3\cdot\frac{2}{9} = \frac{1}{3}$. (16) The cevian-to-a-point lemma (subtract the equal base-ratio pairs sharing the foot and the apex P) gives $\frac{[ABP]}{[ACP]} = \frac{BD}{DC} = \frac{1}{2}$ and $\frac{[ACP]}{[BCP]} = \frac{AE}{EB} = 3$. Set $[BCP] = s$; then $[ACP] = 3s$ and $[ABP] = \frac{1}{2}[ACP] = \frac{3}{2}s$, and since the three pieces cover the triangle, $s + 3s + \frac{3}{2}s = \frac{11}{2}s = 1$. Hence $[APC] = 3s = \frac{6}{11}$.

CHAPTER SIXTEEN

16

Counting Past the Mirror

When two arrangements are "the same" after a flip or a turn, naive counting over-counts. The cure is to count what symmetry leaves untouched.

Burnside: average what stays fixed.

A simple craft project hides a surprising subtlety.

WONDER THIS

You're making a 4-bead *necklace* in a circle, each bead black or white. Rotations count as the same — turn the necklace and it isn't new — but for now we do *not* flip it over. (Once flips count too, the object earns a different name, a *bracelet*; we get there later in the chapter.)

How many genuinely different necklaces are there?

Your first instinct might be $2^4 = 16$, since each of the four beads is black or white. But that counts *arrangements in fixed positions*, not necklaces. The necklace "black, white, white, white" looks identical to "white, black, white, white" after a quarter-turn: they're the same physical object. So 16 badly over-counts. If you carefully list the truly distinct necklaces, you'll find only 6: all-white, all-black, one-black, one-white, two-black-adjacent, and two-black-opposite.

But listing is hopeless for larger necklaces, and the overcounting isn't uniform: the all-black necklace has no rotational copies, while "one black bead" has four. We need a principled way to count objects that are considered "the same" under a family of motions. That family (here, the four rotations of the square) is a *symmetry group*, and the tool that counts correctly is one of the most elegant results in all of mathematics.

CURIOUS CARL • FIRST GUESS

Carl: "I've got it — there are $2^4 = 16$ ways to color the beads, and 4 rotations spin them into each other, so I just *divide*: $16/4 =$ *four* necklaces!"

The catch: nice reflex, Carl — but dividing by $|G|$ pretends every necklace has exactly four rotated copies, and the all-black one has none. The fix isn't to divide; it's to *average* how many colorings each turn leaves unchanged.

THE MOVE

When objects are considered equivalent under a set of **transformations** (rotations, reflections, and so on), the transformations form a *group* G: you can compose any two and get another, every one has an inverse, and there's a do-nothing identity. To count objects *up to* this equivalence, you cannot just divide the naive count by $|G|$ — because different objects have different amounts of symmetry. The correct count requires looking at what each transformation *fixes*.

That last sentence is the crux. The reason "divide by $|G|$" fails is that symmetric objects (like the all-black necklace) are their own rotational copies, so they don't shrink by the full factor. The fix is subtle and beautiful: instead of dividing, we *average the number of arrangements each transformation leaves unchanged.* That's **Burnside's lemma**, and it's the heart of this chapter.

Symmetry groups: cataloguing the motions

Before counting, we list the symmetries. For a square *necklace*, the rotations are by $0°, 90°, 180°, 270°$: four motions forming the *cyclic group* C_4. If we also allow flipping it over, we add four reflections, giving the eight-element *dihedral group* D_4. Now counted up to flips too, the object becomes a *bracelet*. Choosing which symmetries count as "the same" is part of the problem: a bracelet you can flip has more symmetries than a necklace fixed face-up.

WATCH THIS #1 Reading the symmetries of a square

The rotation group C_4. Four elements:

- the identity (turn by $0°$) — leaves every arrangement unchanged;
- rotation by $90°$ and by $270°$ — cycle all four positions around;
- rotation by $180°$ — swaps opposite positions in two pairs.

What each fixes (for 2-color beads). An arrangement is "fixed" by a rotation if it looks identical after the turn:

identity fixes $2^4 = 16$	$90°, 270°$ fix 2 each	$180°$ fixes $2^2 = 4$

The 90° turn cycles all four beads into one orbit, so to look unchanged *all four must match* — only 2 such arrangements (all black, all white). The 180° turn pairs the beads into two opposite pairs, so each pair must match: $2^2 = 4$ arrangements. ■

Trigger Counting configurations that are considered identical under rotations, reflections, or other symmetries — necklaces, colorings, dice, molecular structures.

Do first List the symmetry group, then for each transformation count the arrangements it leaves unchanged (positions forced equal within each cycle of the transformation).

Trapdoor Trying to "divide the total by the group size." That only works when no object has extra symmetry — which is rarely true. Use the fixed-point average instead.

Upgrade The count of arrangements a transformation fixes is $c^{(\text{number of cycles})}$ for c colors — so decompose each transformation into cycles.

The key computational insight, flagged in the upgrade above: a transformation's fixed arrangements are determined by its *cycle structure*. Break a transformation into cycles of positions it permutes; within each cycle every position must wear the same color, so with c colors the number of fixed arrangements is c raised to the number of cycles. The 90° rotation is a single 4-cycle (one cycle, $2^1 = 2$ fixed); the 180° is two 2-cycles ($2^2 = 4$ fixed). Cycle-counting turns symmetry into arithmetic.

△ **TRAPDOOR**

The seductive wrong answer is "total arrangements divided by group size" — here $16/4 = 4$, which is *not* the correct 6. This fails because highly symmetric arrangements (all-black, all-white) don't have a full set of distinct rotational copies, so they're under-divided. Never divide by $|G|$ unless you've checked that the group acts without fixing any nontrivial arrangement. Use Burnside's averaging instead — it automatically handles the symmetric cases.

CURIOUS CARL • TRAPDOOR!

Carl: "Sixteen colorings, four rotations — so $16/4 = 4$ necklaces. Dividing by the group size has to work."

The catch: it works only when every object has a *full* set of four distinct rotated copies. The all-black necklace has just *one* copy — every turn maps it back to itself — so dividing by four strips it of three copies it never had. That is exactly why Burnside *averages* fixed-points instead of dividing: the symmetric arrangements, the ones that break naive division, are precisely the ones the average counts correctly.

The showpiece: Burnside's lemma

Now the theorem that makes it all work: a formula of striking simplicity that correctly counts orbits under any symmetry group.

WONDER THIS

We have all the pieces: a group of symmetries, and for each one, a count of the arrangements it leaves unchanged.

How do these fixed-counts combine into the number of truly distinct objects? The answer is just...their average.

WATCH THIS **#2 Burnside's lemma**

The lemma. The number of distinct objects, counting arrangements as the same when a symmetry relates them, is the *average number of arrangements fixed* by the elements of the group:

$$\#\{\text{distinct objects}\} = \frac{1}{|G|} \sum_{g \in G} |\text{Fix}(g)|,$$

where $\text{Fix}(g)$ is the set of arrangements unchanged by the transformation g.

The necklace, finished. For the 4-bead, 2-color necklace under C_4, we sum the fixed-counts $16 + 2 + 4 + 2 = 24$ and divide by the group size 4:

$$\frac{16 + 2 + 4 + 2}{4} = \frac{24}{4} = 6,$$

exactly matching our hand-count. *The averaging absorbed the uneven overcounting automatically.*

Why it works (the idea). Count pairs $(g, \text{arrangement})$ where g fixes the arrangement, in two ways. Summing over g gives $\sum_g |\text{Fix}(g)|$. Summing over arrangements, each object's orbit contributes exactly $|G|$ to the total (a consequence of how stabilizers and orbits trade off). Equating the two counts and dividing by $|G|$ yields the lemma — it's a double-counting argument, the very technique that opened this book. ■

Burnside's lemma is a showpiece because it dispatches problems that look genuinely hard. Take a cube whose six faces are each painted black or white: how many distinct cubes, accounting for the 24 rotations of the cube? Listing is hopeless. But Burnside reduces it to cycle-counting the 24 rotations: the identity fixes 2^6; the six 90° face-rotations fix 2^3 each; the three 180° face-rotations fix 2^4 each; the eight 120° vertex-rotations fix 2^2 each; the six 180° edge-rotations fix 2^3 each. Averaging,

$$\frac{1 \cdot 2^6 + 6 \cdot 2^3 + 3 \cdot 2^4 + 8 \cdot 2^2 + 6 \cdot 2^3}{24} = \frac{240}{24} = 10.$$

There are exactly 10 distinct two-colored cubes, found not by listing but by averaging fixed-points. *Symmetry, which created the counting problem, also solved it.*

Rotation type	# of rot.	fixed = 2^{cyc}	product
identity	1	$2^6 = 64$	64
face $\pm 90°$	6	$2^3 = 8$	48
face 180°	3	$2^4 = 16$	48
vertex $\pm 120°$	8	$2^2 = 4$	32
edge 180°	6	$2^3 = 8$	48
total = 240 $\lvert G \rvert = 24$ answer = 240/24 =			10

Burnside's lemma on the cube's face 2-colourings: the fixed counts total 240, and $240/24 = 10$.

THE TURN *from grind to insight*

The obvious assault is to write out all $2^6 = 64$ painted cubes, then crawl through them rotating each one 24 ways to weed out the duplicates. The bookkeeping is brutal and one slip ruins the count — some cubes have many copies, some have almost none, and you must track every overlap by hand. **The creative turn** is to list *nothing*: ask only how many colorings each of the 24 rotations leaves *fixed*, then average those numbers, and Burnside's lemma hands back the count of genuinely distinct cubes. It wins because the messy, uneven overcounting — the very thing that wrecks the brute-force tally — is exactly what the averaging dissolves automatically.

Why averaging works: the orbit–stabilizer see-saw. Behind Burnside sits one balancing fact. Every object has an *orbit* (all the arrangements the symmetries can turn it into) and a *stabilizer* (the transformations that leave it unchanged), and the two trade off exactly: $|\text{orbit}| \times |\text{stabilizer}| = |G|$. A highly symmetric object (the all-black necklace) has a *tiny* orbit but a *huge* stabilizer (every rotation fixes it); a lopsided object has a *huge* orbit but a *tiny* stabilizer. Dividing by $|G|$ blindly assumes every object has a full-size orbit, and the symmetric ones don't. That is exactly why it overcorrects them. Averaging the fixed-points instead counts each object's *stabilizer*, and the see-saw guarantees those tally up to give every orbit weight exactly 1. *Don't divide by the symmetries — average what they leave fixed.*

TRY TO BREAK IT

"Divide by $|G|$" works perfectly whenever every object has the same number of rotated copies — so go hunt for the object that breaks it. On the 4-bead necklace, find the coloring with *fewer* than four copies (the all-one-color one is the culprit), and watch how Burnside's average quietly pays it back: the identity over-counts it by 16, but the other three rotations *also* fix it, and those extra fixed-points are exactly the repair that "16/4" was missing.

SEEN BEFORE?

Burnside's proof is a **Chapter 1** double-count (tally the fixed pairs two ways), and the cycle-structure idea connects to the permutation parity of **Chapter 3**. The cube's 24 rotations are the same symmetry group hinted at in Chapter 1's Platonic-solids count — here we use the group to *count colorings* rather than to count the solid's parts.

Payoff: coloring a triangle's corners

A quick one to cement the method.

WONDER THIS

Color the three corners of an equilateral triangle, each corner one of 3 available colors. Two colorings are the same if a rotation of the triangle turns one into the other.

How many distinct colorings are there?

The rotation group is C_3: the identity, a 120° turn, and a 240° turn. Count what each fixes with 3 colors. The identity fixes all $3^3 = 27$ colorings. Each nontrivial rotation is a single 3-cycle of the corners, so all three corners must match: 3 fixed colorings each (one per color). Burnside averages:

$$\frac{27+3+3}{3} = \frac{33}{3} = 11.$$

There are 11 **distinct corner-colorings** of the triangle under rotation. *Eleven, not the naive* $27/3 = 9$ *— the monochromatic colorings, fixed by every rotation, refused to be divided away, and Burnside knew it.* The lemma quietly accounts for every symmetric special case that a careless division would mangle.

SEEN BEFORE?

Burnside upgraded: Pólya enumeration. Burnside counts *how many* distinct colorings exist; it cannot tell you how many use exactly two blacks and one white, say. *Pólya's enumeration theorem* is the weighted generalization that can. Instead of counting each fixed coloring as a plain 1, you attach a weight (a variable per color) and average a polynomial — the *cycle index* of the group, with each color-variable's powers summed inside. Setting every variable to 1 recovers Burnside exactly; keeping them as variables hands you a generating function whose coefficients count colorings by their color content. It is the tool of choice for "how many necklaces with exactly k red beads," and the workhorse behind counting isomers in chemistry and graphs up to symmetry.

WATCH THIS **#3 Pólya in action: necklaces with a fixed palette**

The question. Of the 6 distinct 4-bead necklaces under C_4, how many use *exactly* two red and two blue beads? Plain Burnside can't slice its answer by palette — but watch the weighted version do it.

Attach a marker. Give red the variable r and blue the variable b. Now a bead isn't just "fixed" — it carries its color as a power. When a rotation g groups the four positions into cycles, every position in a cycle must share one color, and that color's marker gets raised to the cycle's length. So a cycle of length k contributes the factor $r^k + b^k$ (all-red or all-blue along that cycle). This is exactly the *cycle index* of C_4 with each a_k replaced by $r^k + b^k$:

$$Z_{C_4} = \tfrac{1}{4}(a_1^4 + a_2^2 + 2a_4) \longrightarrow \tfrac{1}{4}\left[(r+b)^4 + (r^2+b^2)^2 + 2(r^4+b^4)\right].$$

Read off the coefficient. The number of necklaces with two reds and two blues is the coefficient of r^2b^2. Term by term: $(r+b)^4$ gives $\binom{4}{2} = 6$; $(r^2+b^2)^2 = r^4 + 2r^2b^2 + b^4$ gives 2; and $r^4 + b^4$ has no r^2b^2 term at all, so 0. Averaging,

$$[\,r^2b^2\,] = \frac{6+2+2\cdot 0}{4} = \frac{8}{4} = 2.$$

Exactly two such necklaces — the two reds adjacent, or the two reds opposite. *The marker tracked the colors all the way through the average, and Burnside handed back a generating function instead of a plain count.* Set $r = b = 1$ and the whole polynomial collapses to $\frac{16+4+2\cdot 2}{4} = 6$: the total necklace count returns, palette forgotten. This is the same coefficient-extraction game as Chapter 8 — only now the group does the averaging first.

What to try when you're stuck

When a counting problem treats arrangements as identical under rotations, reflections, or other symmetries:

1. **Identify the symmetry group** G and list its elements (rotations, reflections, etc.).
2. **Resist dividing by** $|G|$**.** That overcounts symmetric objects; use Burnside instead.
3. **For each transformation, count fixed arrangements** via its cycle structure: c colors raised to the number of cycles.
4. **Average the fixed-counts:** sum them and divide by $|G|$.
5. **Sanity-check small cases** by hand to make sure your group and cycle counts are right.

THE MOVE

One-page recap. To count objects up to symmetry, average the fixed-points. Toolbox: **identify the group** (cyclic C_n for rotations, dihedral D_n with reflections, the 24 rotations of the cube), **cycle-count each transformation** (fixed arrangements $= c^{\#\text{cycles}}$), and **Burnside's lemma** $\#\text{orbits} = \frac{1}{|G|}\sum_g |\mathrm{Fix}(g)|$. Never just divide by $|G|$ — symmetric objects break it. The averaging handles them automatically.

THE NAME FOR IT

Burnside's lemma

The number of arrangements distinct under a group of symmetries is the average number each symmetry leaves fixed — *Burnside's lemma* (the orbit-counting theorem). The symmetries form a *group*, and the bridge from fixed points to orbit sizes is the *orbit–stabilizer theorem*.

WHERE IT LEADS Burnside's lemma is the doorway to *group representation theory* and *Pólya enumeration*: counting orbits by averaging over a group is exactly how characters and the *orbit-counting* machinery of physics and chemistry (molecular symmetry, crystallography) work. Groups acting on sets are the heart of modern *symmetry*.

CURIOUS CARL • EUREKA

Carl: "I ask each turn *what did you leave alone?* — sixteen, two, four, two — then *add* them. Twenty-four necklaces!"

Almost — the last inch: those four numbers are exactly right, Carl. Now *average* them instead of adding: Burnside divides by the number of turns, so $24/4 = 6$. One word — "average," not "add" — and you'd have had the whole theorem.

SAME BEAST, NEW COSTUME

The same "average what stays fixed" beast, dressed up across the book:

Necklace counts → Cube colorings → Burnside's lemma → The $\varphi(d)$ necklace formula.

Each one refuses to be counted by naive division, because symmetric objects hold still under too many motions — so you average the fixed-points and let the symmetry pay its own bill.

The practice ladder

THE PRACTICE LADDER *climb at your own pace*

LAUNCH *Warm the engine. Nudges below.*

1. List all distinct 3-bead, 2-color necklaces under rotation, then confirm with Burnside.
2. How many arrangements does a 180° rotation of a 6-bead necklace leave unchanged, with 2 colors? (Find its cycle structure.)

CORE BUILDER *Master the central moves.*

1. Count the distinct 6-bead, 2-color necklaces under rotation (C_6) using Burnside.
2. Count the distinct 4-bead, 2-color bracelets under the full dihedral group D_4 (rotations *and* reflections — so a coloring and its mirror image count as the same), and compare with

the rotation-only count of 6.

3. How many distinct ways to color the 4 corners of a square with 3 colors, under rotation?

STRETCH *Transfer to unfamiliar ground.*

1. Count the distinct ways to color the 6 faces of a cube with 3 colors under the 24 rotations.

2. Count the distinct ways to color the edges of an equilateral triangle with 2 colors under the dihedral group D_3.

THE FLASH *No clock — each cracks open in one flash of insight, once you find the angle. Nudges at the back.*

1. How many distinct 2-colored cubes (faces) are there under rotation?

2. How many distinct 3-color corner-colorings of a triangle under rotation?

PROOF LAB *Write it clean, in full sentences.*

1. Prove Burnside's lemma by counting fixed pairs (g, x) with $g \cdot x = x$ in two ways, invoking the orbit–stabilizer relationship.

2. Derive the necklace-counting formula $\frac{1}{n}\sum_{d|n}\varphi(d)\,c^{n/d}$ for n-bead, c-color necklaces under rotation, connecting the Euler totient (Chapter 10) to the cycle structure of rotations.

REFLECTION *No nudge. Write a few sentences.*

1. For one problem above, explain why dividing the naive count by $|G|$ gives the wrong answer, and identify which symmetric arrangements caused the discrepancy.

NUDGES

Launch. (1) $\frac{2^3+2+2}{3} = 4$ distinct necklaces. (2) The 180° turn is three 2-cycles, fixing $2^3 = 8$. **Core.** (3) $\frac{1}{6}(2^6 + 2 + 2^2 + 2^3 + 2^2 + 2) = \frac{84}{6} = 14$. (4) D_4: rotations fix 16, 2, 4, 2; reflections fix 8, 8, 4, 4; total $48/8 = 6$ (here equal to rotation count). (5) $\frac{1}{4}(3^4 + 3 + 3^2 + 3) = \frac{96}{4} = 24$. **Stretch.** (6) $\frac{1}{24}(3^6 + 6 \cdot 3^3 + 3 \cdot 3^4 + 8 \cdot 3^2 + 6 \cdot 3^3) = \frac{1368}{24} = 57$. (7) Edges of a triangle under D_3: $\frac{1}{6}(2^3 + 2 \cdot 2 + 3 \cdot 2^2) = \frac{24}{6} = 4$. **The Flash.** (8) 10. (9) 11. **Proof.** (10) Count the pairs (g, x) with $g \cdot x = x$ two ways: by group elements it is $\sum_g |\text{Fix}(g)|$, by points it is $\sum_x |\text{Stab}(x)|$. Orbit–stabilizer gives $|\text{Stab}(x)| = |G|/|\text{orbit}(x)|$, so the points of a single orbit contribute $|\text{orbit}(x)| \cdot \frac{|G|}{|\text{orbit}(x)|} = |G|$; hence $\sum_x |\text{Stab}(x)| = |G| \cdot (\#\text{orbits})$. Equating the two counts gives $\#\text{orbits} = \frac{1}{|G|}\sum_g |\text{Fix}(g)|$. (11) Rotation by k splits the n beads into $\gcd(k, n)$ cycles, and a coloring is fixed iff it is constant on each cycle, so $|\text{Fix}(r^k)| = c^{\gcd(k,n)}$. Burnside then gives $\frac{1}{n}\sum_{k=0}^{n-1} c^{\gcd(k,n)}$; grouping the k by $d = \gcd(k, n)$, exactly $\varphi(n/d)$ values of k produce each divisor d, so the sum becomes $\frac{1}{n}\sum_{d|n}\varphi(n/d)\,c^d$. Reindexing $d \mapsto n/d$ yields $\frac{1}{n}\sum_{d|n}\varphi(d)\,c^{n/d}$.

Chapter Seventeen

17

Dots, Lines, and Lonely Hearts

Strip away everything but who-connects-to-whom, and a surprising amount of the world — one-stroke doodles, impossible maps, even who can marry whom — turns out to be dots and lines.

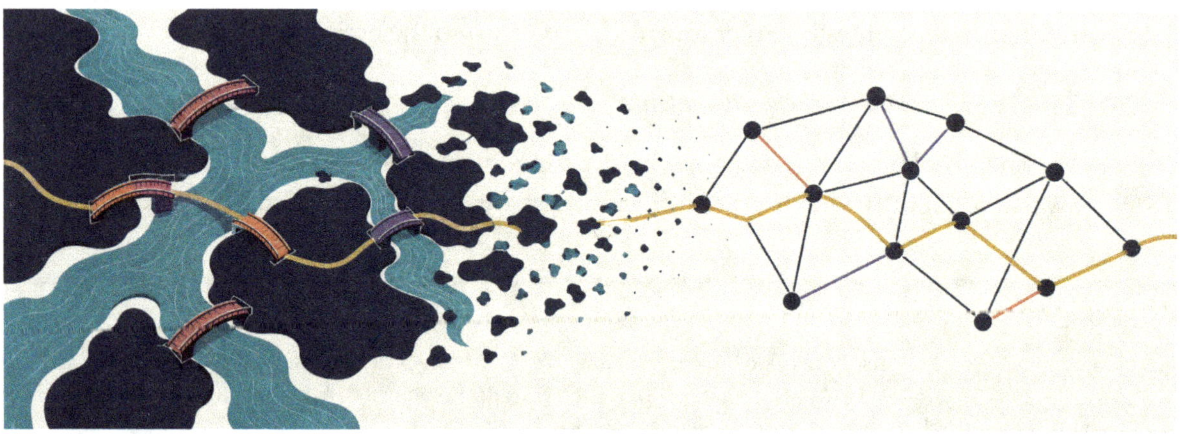

Graph theory: count locally, conclude globally.

Everyone has attempted the doodle below.

WONDER THIS

Can you draw the following figure *without lifting your pen* and without tracing any line twice? Picture the sealed-envelope doodle: a rectangle with both of its diagonals drawn in and a triangular flap on top — or try the classic "Bridges of Königsberg," four landmasses joined by seven bridges, walked so each bridge is crossed exactly once.

Which such figures can be drawn in one stroke, and which can't — and how can you tell at a

glance?

You can spend a frustrating afternoon trying and failing on some of these. The breakthrough (due to Euler, and often called the birth of graph theory) is to stop thinking about the drawing and think only about the *structure*: which points connect to which. Reduce the figure to *vertices* (the dots where lines meet) and *edges* (the lines between them). Everything about "can I draw it in one stroke" depends on a single, easily-checked feature of that structure.

CURIOUS CARL • FIRST GUESS

Carl: "Can five people each shake exactly three hands? Let me *draw* it — five dots, and I'll try to hang three lines off each until it works!"

The catch: put the pen down, Carl — no drawing needed. Add up the degrees: $5 \times 3 = 15$, but the degrees of any graph sum to $2E$, which is always *even*, so 15 is flatly impossible. One local count kills it before you draw a single line.

The feature is the *degree* of each vertex: how many edges meet there. Here's Euler's insight: as your pen travels through a vertex in the middle of the journey, it uses up *two* edges, one to arrive, one to leave. So every "pass-through" vertex needs an even degree. Only the start and end of your stroke can have an odd degree (the start has an unmatched leaving-edge, the end an unmatched arriving-edge). Therefore:

THE MOVE

Model a connected figure as a **graph** of vertices and edges. A single unbroken stroke tracing every edge exactly once — an **Euler trail** — exists if and only if the number of *odd-degree* vertices is *exactly* 0 *or* 2. If 0, you can even return to your start (an Euler circuit); if 2, you must start at one odd vertex and end at the other; if more than 2, it's impossible. (Connectedness is the fine print: isolated dots may be ignored, but every edge must lie in a single connected piece — two separate loops have zero odd vertices yet plainly cannot be drawn in one stroke.) The drawing question became a degree-counting question.

That's the whole answer. The Königsberg bridges fail because all four landmasses have odd degree $(3, 3, 3, 5)$: four odd vertices, far more than two, so no single-stroke walk exists, and centuries of towns-folk had been attempting the impossible. The envelope works precisely when it has zero or two odd corners. **Count the odd-degree vertices; the answer is immediate.** This is the spirit of graph theory: abstract away the irrelevant, and a hard-looking problem reduces to counting a simple local feature.

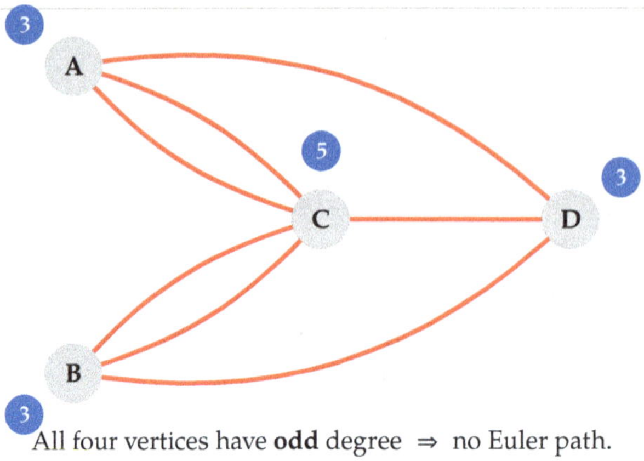

The seven bridges as a multigraph: all four land masses have odd degree, so no Euler walk can exist.

The handshake lemma: a tiny law with big reach

Before going further, here is the most fundamental fact about **graphs**. It is so simple it's almost a joke, yet it proves things you'd never expect.

WATCH THIS — **#1 The handshake lemma**

Claim. In any graph, the sum of all the vertex degrees equals *twice* the number of edges — and so the number of odd-degree vertices is always *even*.

Why. This is exactly the **Chapter 1** double count in graph clothing: tally the (vertex, incident-edge) pairs two ways. By vertices the total is $\sum_v \deg(v)$; by edges it is $2E$, since each edge has two ends. Hence $\sum_v \deg(v) = 2E$, an even number. The even-degree vertices contribute an even amount, so the odd-degree vertices must come in an even count. ■

Trigger Any statement about degrees, edge counts, or "can everyone have such-and-such many connections," including party/handshake puzzles.

Do first Apply the handshake lemma: $\sum \deg = 2E$, and the count of odd-degree vertices is even.

Trapdoor Forgetting that the lemma counts each edge twice — the factor of 2 is the whole point and the source of the "even" conclusions.

Upgrade This instantly answers "can n people each shake exactly k hands" (need nk even) and explains why the Euler-trail count of odd vertices is 0 or 2, never 1.

This little lemma is pure Chapter 1 double-counting, and it pays off everywhere. It explains why an Euler trail can't have exactly one odd vertex (odd vertices come in pairs). It tells you instantly that you cannot have a party of five people where everyone shakes exactly three hands (that would need $5 \times 3 = 15$ edge-endpoints, an odd number — impossible). The lesson: *in graphs, counting the same thing two ways (once per vertex, once per edge) is the master key.*

THE TURN *from grind to insight*

The head-on attempt at "can five people each shake exactly three hands" is to start drawing the network and case-analyze the whole tangle — who pairs with whom, branching on every choice until something jams. The cases multiply and you can wander for a long time before the contradiction surfaces. **The creative turn** is to ignore the global structure entirely and tally a single local quantity: add up the degrees. That sum equals $2E$, hence is always *even* — but five threes make 15, odd, so no such graph can exist. It wins because one clean count of a local feature settles a global question that brute-force casework can only grind at.

TRY TO BREAK IT

The handshake lemma is only the *first* filter. An even degree-sum is *necessary* but not *sufficient*: can a simple graph have degrees $4, 4, 3, 1, 1, 1$? The sum is 14, even, so the lemma is content — yet no such graph exists. The detective's tool is the **Havel–Hakimi** reduction: sort the degrees largest-first, delete the top degree d, and subtract 1 from the next d entries (that vertex spends its d edges on the hungriest others); repeat. Reach all zeros and the list is *graphical*; run a subtraction off the end or into a negative and it is impossible. On $4, 4, 3, 1, 1, 1$ the leading 4 leaves $3, 2, 0, 0, 1$; re-sorted to $3, 2, 1, 0, 0$, the leading 3 must now subtract from a 0, going negative — so the list can never be drawn, however you try.

Degrees hold one more party trick, this time powered by pigeonhole rather than parity.

WATCH THIS **#2 The party twins**

Claim. At any party with $n \geq 2$ guests, some two guests have exactly the same number of friends present.

Why. Each guest's friend-count lives in $\{0, 1, \ldots, n-1\}$ — apparently n possible values for n guests, so no collision looks forced. But the two extremes cannot coexist: a *hermit* (count 0) and a *social butterfly* (count $n-1$, friends with everyone else) would have to be friends and strangers at once. So at most $n-1$ of the values actually occur, and the pigeonhole principle (Chapter 2!) forces two guests to share one. No matter how the friendships fall, the party always contains degree-twins. ■

△ TRAPDOOR

"Degree" counts edge-*ends* at a vertex, not neighbors. A self-loop (an edge from a vertex to itself) contributes 2 to that vertex's degree, and multiple edges between the same pair each count separately. When a problem allows loops or repeated edges, tally edge-ends carefully — the handshake lemma still holds, but only if you count degree as edge-ends. (Unless a puzzle explicitly involves loops or repeated edges, every graph in this chapter — and in particular each planar edge bound and coloring result below — is assumed *finite* and *simple*.)

Loops are not lawyer's fine print, either. They star in a puzzle sitting in your game cupboard.

WATCH THIS #3 The domino ring

Claim. The 28 tiles of a double-six domino set (every pair from 0–6, doubles included) can always be laid in a single *closed ring*, touching ends matching. Better: if anyone claims to have laid all 28 in an *open* line, peek at the two exposed ends — they must show the *same* number.

Why. Change the model. The numbers 0–6 become seven *vertices*, and each tile $[a \mid b]$ becomes the *edge* joining a to b — a double $[a \mid a]$ is a loop at a. A legal chain, matching ends touching, is a walk using every edge exactly once: *an Euler trail in disguise*. Now count degrees: each number meets the six others once (6 edge-ends) and its own double contributes 2 more — degree 8, **even**, at all seven vertices. The graph is connected, so an Euler *circuit* exists: a closed ring, every time. The same count settles the open-line boast: with all degrees even, a trail using every edge must end where it began — at the same *number* — so any "open" line of all 28 secretly wants to close into the ring. ■

TRY TO BREAK IT

Now sabotage it: discard the seven tiles containing a 6, leaving the 21 tiles of a double-five set, and recount. Each number 0–5 now has degree $5 + 2 = 7$ — *odd*, at all *six* vertices. Six odd vertices is four too many: no chain, open *or* closed, can ever use all 21 tiles. One rank of tiles removed, and the ring is not merely harder — it is provably dead. (A double-nine set fails the same way: ten vertices, each of odd degree 11.)

The showpiece: Euler's formula

Now a result of breathtaking generality: a single equation governing every map, every polyhedron, every network you can draw without crossings.

WONDER THIS

Draw any connected graph on paper so that no two edges cross — a "planar" drawing. It carves the plane into regions, or *faces* (counting the unbounded outside as one face).

Count the vertices V, edges E, and faces F. Is there a relationship among them? Try a square with a diagonal, or a cube's edge-diagram.

Tabulate a few and a pattern leaps out. A triangle: $V = 3, E = 3, F = 2$ (inside and outside), and $3 - 3 + 2 = 2$. A cube drawn flat: $V = 8, E = 12, F = 6$, and $8 - 12 + 6 = 2$. Always 2.

WATCH THIS #4 Euler's polyhedron formula

Claim. For any connected planar graph, $V - E + F = 2$.

Proof by demolition. Start with the graph drawn in the plane and repeatedly simplify, watching $V - E + F$ never change:

- **If there's a cycle,** delete one edge of it. This merges two faces into one, so E drops by 1 and

F drops by 1 — the quantity $V - E + F$ is unchanged.

- **Repeat** until no cycles remain. The graph is now a *tree* (connected, no cycles).

graph: $V - E + F$ —— **strip cycles** ——> **tree:** $V - (V-1) + 1$

$= 2$

The base case. A tree on V vertices has exactly $V - 1$ edges and only $F = 1$ face (no cycles enclose anything). So

$$V - E + F = V - (V - 1) + 1 = 2.$$

Since stripping cycles never changed $V - E + F$, the original graph also had $V - E + F = 2$. ■

CURIOUS CARL • TRAPDOOR!

Carl: "I'll sketch the graph in two separate pieces, plug V and E into $V - E + F = 2$, and read the face-count straight off."

The catch: Euler's $V - E + F = 2$ only speaks for a *connected* (and planar) graph; feed it two disconnected islands and the count drifts to 3, one extra for the piece left stranded. A right-looking V and E never promise a graph is connected *or* drawable flat — check those first.

Euler's formula is a showpiece because of its *reach*: it's a single invariant (there's that word from Chapter 3 again) that holds for every planar graph, and from it cascades an astonishing amount. It proves that the five Platonic solids are the only ones possible (the same fact Chapter 1 reached by double-counting, now from topology). It shows certain graphs simply *cannot* be drawn without crossings. The complete graph K_5 on five mutually-connected points has $E = 10$ edges but only $3V - 6 = 9$, so it breaks the bound $E \le 3V - 6$ that Euler's formula forces on every simple planar graph with $V \ge 3$. The "utilities graph" $K_{3,3}$ (three houses each joined to three utilities) is subtler: with $V = 6$, $E = 9$ it actually *obeys* $E \le 3V - 6 = 12$. But $K_{3,3}$ is *bipartite*, so it contains no triangles, and a triangle-free planar graph must satisfy the sharper bound $E \le 2V - 4$. (Why the sharper bound? In a simple triangle-free planar drawing every face is bounded by at least 4 edges, so counting edge–face incidences gives $4F \le 2E$, hence $F \le E/2$; substituting into $V - E + F = 2$ yields $V - E + \frac{E}{2} \ge 2$, i.e. $E \le 2V - 4$.) For $K_{3,3}$ this reads $E \le 8$; since $9 > 8$, $K_{3,3}$ is nonplanar too. *One little alternating sum, $V - E + F$, constrains the entire geometry of what can be drawn flat.*

THE MOVE

The dual: turn faces into vertices. Every planar drawing hides a second graph. Drop one new dot inside each face and join two of them whenever their faces share an edge — that is the **dual graph**, in which boundaries become crossings and crossings become boundaries. The Platonic solids pair off exactly this way: the cube (six faces, three meeting at a corner) dualizes to the octahedron (six corners, where those faces now meet), and back again; the dodecahedron and icosahedron swap likewise; and the tetrahedron is its own dual. *The graph has a shadow hiding in* *its faces.*

It also settles a question you have been kicking around since childhood.

WATCH THIS **#5 The soccer-ball theorem**

Claim. Stitch a ball from pentagons and hexagons, three seams meeting at every corner (the classic soccer ball does exactly this). However many hexagons you use, **the number of pentagons is always exactly** 12.

Why. Say there are P pentagons and H hexagons, so $F = P + H$. Two Chapter 1 double counts: every corner emits 3 seams and every seam has 2 ends, so $3V = 2E$; and counting (face, bordering-seam) pairs, $5P+6H = 2E$, since each seam borders two faces. Feed both into Euler's $V - E + F = 2$:

$$\frac{2E}{3} - E + P + H = 2 \implies 6P + 6H - 2E = 12$$
$$\implies 6P + 6H - (5P + 6H) = 12 \implies P = 12.$$

The hexagons cancel and vanish from the equation entirely: H can be 20 (the standard ball) or enormous (chemists build giant "fullerene" carbon molecules on the same plan), but the twelve pentagons are non-negotiable — Euler's formula decided before the first stitch. ■

TRY TO BREAK IT

Euler's $V - E + F = 2$ leans on a hidden crutch: the graph must be *connected*. Draw two separate triangles on the same page and recount — you get $V = 6$, $E = 6$, and $F = 3$ (two insides plus the one shared outside), so $V - E + F = 3$, not 2. Where did the extra go? Track it as you glue the pieces together with a single new edge and watch the count snap back to 2; you'll see the "2" was really "1 + (number of pieces)" all along.

TRY TO BREAK IT

One more hypothesis to attack: the *surface*. $K_{3,3}$ cannot be drawn on this flat page — but draw it on a **coffee mug**, and the ninth edge escapes over the handle, crossing nothing. Euler's "2" is a fact about the plane (equivalently, the sphere), not about graphs: on the mug's surface — a sphere wearing one handle — the invariant shifts, and careful drawings give $V - E + F = 0$ instead. That number, the *Euler characteristic*, is the first fingerprint topologists use to tell surfaces apart; the formal-terminology box has the forwarding address.

SEEN BEFORE?

The handshake lemma is a **Chapter 1** double-count, and $V - E + F = 2$ is an **Chapter 3** invariant — a quantity unchanged by the demolition moves. The "strip it down to a base case" demolition is the descent/monovariant idea again, and Euler's formula reproves the Platonic-solids count from Chapter 1 by a completely different route.

The marriage theorem: when can everyone be matched?

Some of the most useful graphs are *bipartite*: two groups, with edges only *between* them, never inside. Think applicants and jobs, students and projects, or — the classic framing that named the theorem — people and the partners they'd be willing to marry.

WONDER THIS

Each of n people has a list of acceptable partners (drawn from a separate pool). We want a *perfect matching*: pair everyone off so that each person gets someone on their list, and nobody is shared.

When is this possible? With 100 people and tangled, overlapping preference lists, must you check every conceivable pairing — or is there a single clean condition that decides it?

There is one condition, and it is exactly what you'd worry about. If some group of 3 people collectively find only 2 partners acceptable among them, they can't all be matched: three into two won't go. Philip Hall's theorem says this obvious obstruction is the *only* one.

WATCH THIS **#6 Hall's marriage theorem**

Claim. A perfect matching exists if and only if *Hall's condition* holds: *every* set S of people collectively lists at least $|S|$ acceptable partners. In symbols, $|N(S)| \geq |S|$ for all S, where $N(S)$ is the set of partners acceptable to someone in S.

The necessary direction is free. If some S has $|N(S)| < |S|$, those $|S|$ people are fighting over fewer than $|S|$ partners — by pigeonhole (Chapter 2!) two must share, so no perfect matching exists. The condition is clearly *necessary*.

The surprise is that it's also sufficient: whenever every group has enough options collectively, a full matching can always be assembled. That direction is the substance of the theorem; its proof is an optional detour below. For solving problems, the working tool is the condition itself — check whether some group is over-subscribed. ∎

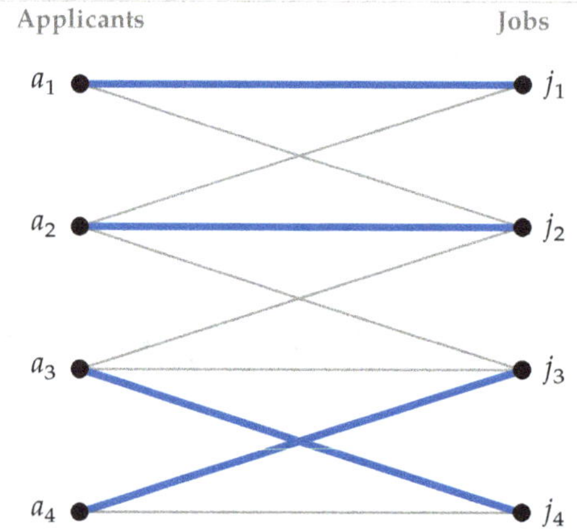

Hall's condition: every set S of applicants reaches $|N(S)| \geq |S|$ jobs $\Rightarrow$ a perfect matching exists.

Hall's theorem is the backbone of *matching theory*: scheduling, assigning tasks to workers, routing data, even the algorithms behind medical-residency placement all rest on it. *The romance is optional — but the lesson is that a single local condition, "no group is over-subscribed," globally guarantees everyone finds a match.*

And it performs actual magic.

WATCH THIS **#7 The card trick**

Claim. Shuffle a standard 52-card deck however you like and deal it into 13 piles of 4. It is *always* possible to pick one card from each pile so that the 13 chosen cards show all 13 ranks — ace through king, one of each.

Why. Build the bipartite graph: the 13 piles on the left, the 13 ranks on the right, with an edge when a pile contains a card of that rank. Take any set S of k piles. Together they hold exactly $4k$ cards; and since each rank supplies only 4 cards to the entire deck, no single rank can account for more than 4 of those cards. So the $4k$ cards must spread across at least k distinct ranks: $|N(S)| \geq k = |S|$. Hall's condition holds *automatically* — for every deal, with no cleverness — so a perfect matching of piles to distinct ranks always exists. A magician calls it luck; Hall calls it a theorem. The shuffle was never given a chance. ■

GOING FURTHER *optional • beyond AIME*

Optional, beyond-contest. The necessary direction above is what you actually use to solve matching problems. The reverse implication — that Hall's condition is *sufficient*, i.e. that it always forces a full matching into existence — is the hard and beautiful half of the theorem. Here is the classic proof for the curious; nothing later in the chapter relies on it.

We argue by induction on n, the number of people, assuming Hall's condition $|N(S)| \geq |S|$ holds for every set S. The base case $n = 1$ is immediate: one person with at least one acceptable partner gets matched. For the inductive step, split into two cases according to how tight the condition is.

Case 1: slack everywhere. Suppose every *proper* non-empty group S (that is, $1 \leq |S| < n$) has strictly more options than members, $|N(S)| > |S|$. Pick any one person and match them to any acceptable partner p. Removing that person and p drops the available-partner count of any remaining group by at most one, so each remaining group S still has $|N(S)| \geq |S|$ — Hall's condition survives. By induction the other $n - 1$ people can all be matched, completing the matching.

Case 2: a tight group. Suppose instead that some proper non-empty group S is exactly tight, $|N(S)| = |S|$. Within S and its neighborhood $N(S)$, Hall's condition holds (any subgroup of S already satisfied it), so by induction S can be perfectly matched to $N(S)$, using up all of $N(S)$. Now consider the leftover people, those outside S. For any group T of them, look at $S \cup T$: by Hall, $|N(S \cup T)| \geq |S| + |T|$, and since $N(S)$ has exactly $|S|$ partners all consumed by S, the partners available to T *outside* $N(S)$ number at least $|T|$. So the leftover people, drawing only on partners outside $N(S)$, satisfy Hall's condition among themselves; by induction they too can be matched. Combining the two matchings pairs everyone off.

Either case produces a perfect matching, completing the induction. ■

Where it leads. This existence theorem is the seed of *matching theory* and the wider study of network flows and combinatorial optimisation; the same idea, run as an algorithm, is how real scheduling and assignment systems actually find their matchings.

The board becomes a graph: dominoes on the chessboard

Half the puzzles in any contest live on a square grid — boards to tile, cells to visit, corners to reach. Here is a modeling move that turns all of them into this chapter's home territory: *make the board a graph.* Each cell becomes a vertex; two cells that share a *side* are joined by an edge. Once the board is a graph, every tool of this chapter — degrees, paths, cycles, matchings, Hall — is suddenly aimed straight at tiling puzzles.

WONDER THIS

An 8×8 chessboard loses two *opposite corner* cells, leaving 62. You hold 31 dominoes, each covering exactly two cells that share a side.

Can the 31 dominoes tile the mutilated board? Sixty-two cells, thirty-one dominoes — the raw count is perfect. Try it, though, and every attempt jams near the end.

THE MOVE

The board-to-graph dictionary. Convert the board: **cells become vertices**, and **two cells that share a side become an edge** — the board's *grid graph.* Under this dictionary, a *domino* is an *edge* (it covers the edge's two endpoints); a *tiling by dominoes* is a *perfect matching* (every vertex covered exactly once, no overlaps); a *walk* through side-adjacent cells is a *path*, and a round trip is a *cycle*. Best of all, the classic checkerboard coloring is exposed as structure: the cells (i, j) with

$i + j$ even and those with $i + j$ odd form the two sides of a *bipartition* — every side-step changes the parity of $i + j$, so every edge joins opposite colors, and **the grid graph is bipartite.**

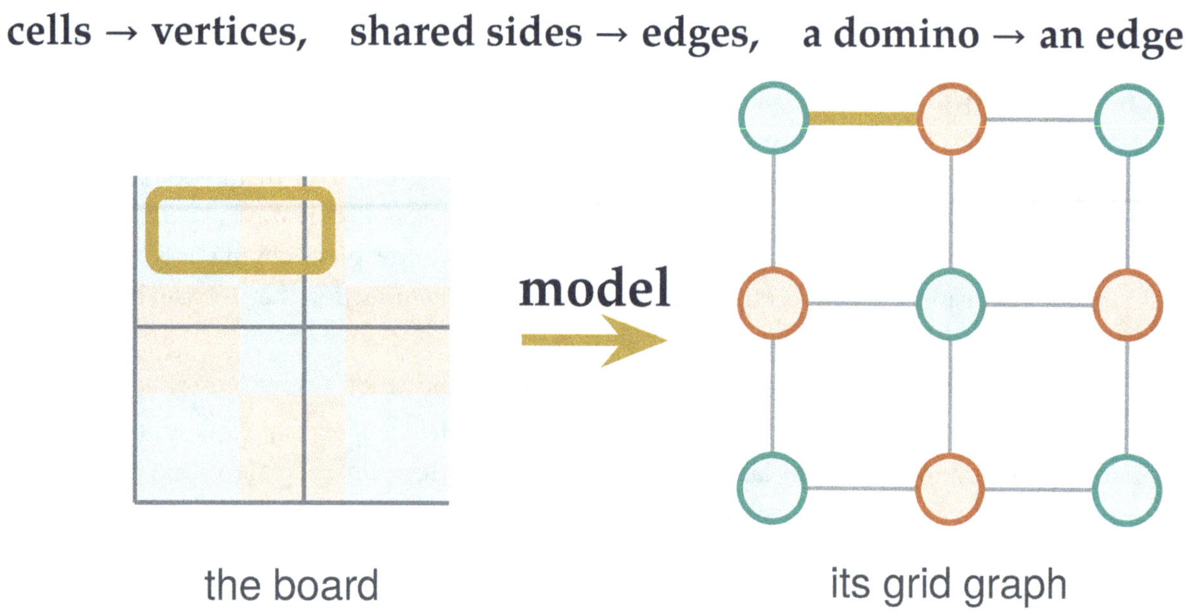

Now the coldopen dissolves. A perfect matching of a bipartite graph pairs the two color classes off one-to-one, so it needs the classes *equal*. But opposite corners of a chessboard have the *same* color: the mutilation removes two cells from one class, leaving 30 of one color facing 32 of the other. **Unequal classes, no perfect matching, no tiling** — and the 31-domino head count never had a say. In Hall's language, take S to be all 32 majority cells: their neighbors all lie among the 30 minority cells, so $|N(S)| = 30 < 32 = |S|$, and Hall's condition dies at the largest possible set.

THE TURN *from grind to insight*

The head-on attempt is to tile by hand and case-split: lay a domino at the damaged corner, branch on its orientation, and chase the tree of layouts until every branch jams — a jungle with thousands of dead ends. **The creative turn** is to convert the board to its grid graph, notice it is bipartite, and compare the two color classes: $30 \neq 32$, so no perfect matching can exist. One subtraction replaces the jungle. It wins because the graph model exposes the board's hidden two-class structure, and a perfect matching is hostage to it.

Do not walk away believing equal color counts *guarantee* a tiling — Hall's condition patrols *every* group, not just the biggest. Remove the two neighbors of one corner cell (both the opposite color), plus any two cells of the corner's own color elsewhere: the classes balance at 30–30, yet the corner is now an *isolated vertex*. The one-cell set $S = \{\text{corner}\}$ has $N(S) = \emptyset$, and the tiling is dead before the count even matters. *The coloring count is Hall's condition applied to the largest set; the full condition also watches every smaller one.*

And here is the converse showpiece. Which removals *always* leave a tileable board? Cycle thinking answers it.

WATCH THIS **#8 Gomory's theorem**

Claim. Remove any *one* black cell and any *one* white cell from the full 8×8 board. The remaining 62 cells can *always* be tiled by 31 dominoes.

Why. The grid graph of the board contains a *closed tour* through all 64 cells by side-steps — a Hamiltonian cycle: snake up one column and down the next, then return home along the bottom row (the picture below shows the pattern on a 4×4 board). Walking the cycle, the colors alternate black, white, black, white, all the way around. Now delete one black cell and one white cell: they cut the cycle into two arcs, and because the colors alternate, *each arc contains an even number of cells.* (If the two removed cells happen to be neighbors on the tour, one "arc" is empty — an even length of zero, which causes no trouble.) Lay dominoes along each arc, two consecutive cells at a time: consecutive cells of the tour share a side, so every domino is legal, and both arcs are tiled completely. ■

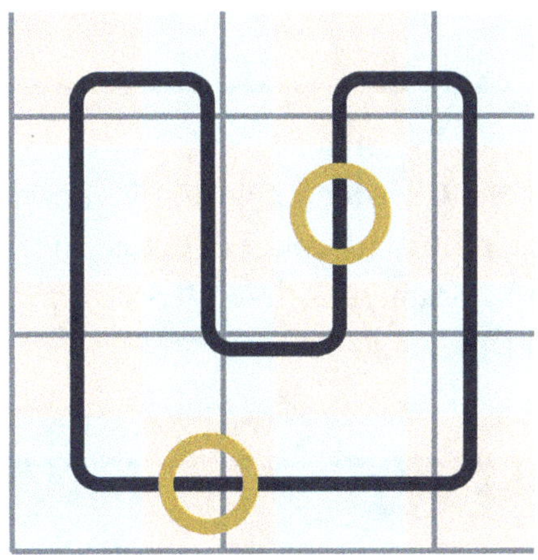

a closed side-step tour of every cell; removing one cell of each color (circled) cuts it into two even arcs

The same dictionary powers movement puzzles. A beetle wants to crawl through every cell of a 5×5 board exactly once, always stepping to a side-adjacent cell. In graph terms that is a *Hamiltonian path* in the grid graph — and bipartiteness instantly constrains it. The path's 25 cells alternate colors with every step, reading (corner color), (other), (corner color), ...; with $25 = 13 + 12$, it uses thirteen cells of the corner color and twelve of the other, and *both* of its endpoints land in the thirteen. **Any such crawl must start and end on cells of the corner color** — so if a problem demands a tour starting beside a corner, it is over before it begins. The two-coloring that looked like a bolt of cleverness in earlier chapters stands revealed as the bipartite structure of one graph. (And why must the two-coloring exist at all? Because the grid graph has no *odd cycle*: any round trip returns to its starting color, so its length is even — and "bipartite" is exactly "no odd cycles.")

△ **TRAPDOOR**

The dictionary says *side*, not corner: diagonally touching cells are **not** adjacent in the grid graph, and a domino never covers a diagonal pair. And the graph depends on the move rule — a king, a knight, and a rook each generate a *different* graph on the same 64 cells (a knight's tour is a Hamiltonian cycle in the *knight-move* graph, bipartite for the very same coloring reason). Always build the graph from the problem's actual adjacency before you count anything.

Trigger Boards, domino and tromino tilings, beetle or rook tours, "pair up the cells" puzzles — anything that lives on a grid of cells.

Do first Convert the board into its grid graph (cells → vertices, shared sides → edges); tilings become perfect matchings (compare the color classes, then check Hall), and tours become paths or cycles (use the alternation of the bipartition).

Trapdoor
Stopping at the two color counts — equal counts do *not* guarantee a tiling, because Hall's condition can still fail at a small local set (an isolated or starved corner).

Upgrade The mutilated chessboard falls in one line ($30 \neq 32$); Gomory's theorem falls to one Hamiltonian cycle; and every "visit each cell once" puzzle surrenders its endpoints to the bipartition.

SEEN BEFORE?

The checkerboard coloring is the **Chapter 3** coloring-invariant idea, now unmasked as the *bipartition* of the grid graph; the mutilated-board count is **this chapter's** Hall condition applied to its largest set; and Gomory's argument is the Euler-trail spirit again — build one grand walk first, then harvest the answer from its structure.

GOING FURTHER *optional • beyond AIME*

Optional, a look over the fence. This chapter's two kinds of grand tour look like twins and could not be more different. A tour through every *edge* exactly once is an *Euler trail*, and we settled it completely with a glance at the degrees. A tour through every *vertex* exactly once is a *Hamiltonian path* (a *Hamiltonian cycle* if it returns home) — named for William Rowan Hamilton, who in 1857 sold it as a commercial puzzle, the *Icosian game*, played on the corners of a dodecahedron. Here is the shock: **no simple test for Hamiltonian tours is known.** Not for lack of trying — deciding whether a graph has one is a template "hard problem" of modern computer science: a *proposed* tour can be checked in moments, but no known method *finds* one (or certifies that none exists) essentially faster than brute search. Whether checking-is-easy must imply finding-is-easy is the celebrated *P versus NP* question, with a million-dollar bounty on its head. Notice how we handled our own Hamiltonian moments: Gomory's cycle and the beetle's path were never found by a test — we *built* the cycle by hand and *constrained* the path by bipartite counting. Against Hamilton, construction and obstruction are all anyone has; against Euler, a degree count is a complete theory. Knowing which kind of tour a problem asks for tells you which kind of fight you are in.

The second modeling move: every puzzle is a graph

The grid graph turned a board's *space* into a graph. The chapter's second modeling move turns a puzzle's *unfolding in time* into one. It swallows river crossings, jug pourings, sliding blocks, and the Rubik's Cube whole.

WONDER THIS

A farmer must ferry a wolf, a goat, and a cabbage across a river. The boat holds the farmer plus one passenger. Left alone together without the farmer, the wolf eats the goat, and the goat eats the cabbage.

Can everyone cross intact — and what is the smallest number of crossings?

THE MOVE

The state graph. Make every legal *position* of a puzzle a vertex, and join two positions by an edge when one legal *move* turns either into the other. Then a *solution* is a *path* from the start vertex to the goal vertex; the *fastest* solution is the graph *distance* between them; and the hardest legal starting position is the vertex *farthest* from the goal. Solving stops being cleverness and becomes geography.

For the river puzzle, a position records who stands on which bank. Not all $2^4 = 16$ assignments are legal — the wolf–goat and goat–cabbage vetoes strike whenever the farmer is absent — and exactly 10 survive. Draw the ten, join positions one boat-trip apart, and the entire puzzle appears at a glance:

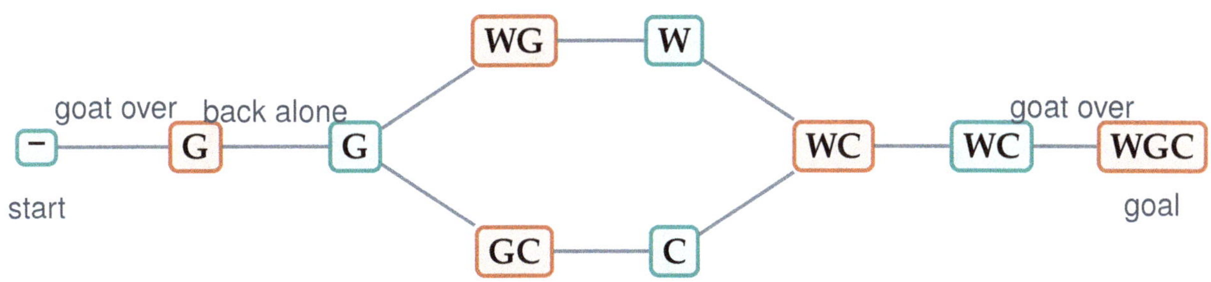

each box lists what has crossed so far; teal = farmer on the near bank, coral = farmer across — the solution is a 7-edge path, and the diamond records the one free choice

Every question about the puzzle dissolves into reading the map. Is it solvable at all? Yes — the graph is *connected*. What is the fastest solution? The distance from start to goal is 7: seven crossings, and provably no fewer. Are there genuinely different solutions? Exactly two shortest ones: the diamond in the middle records the single free choice (ferry the wolf across second, or the cabbage), and the graph shows the two routes are mirror images of each other. **The state graph doesn't just solve the puzzle; it exhibits every solution at once, and certifies the minimum.**

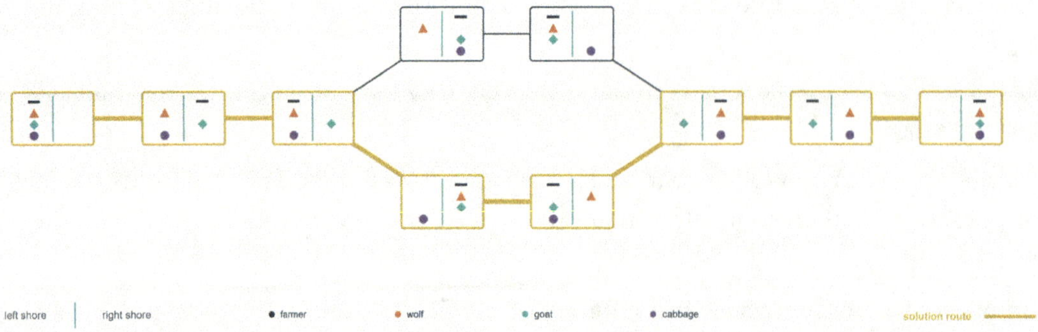

Puzzle designers call the distance from the farthest position a puzzle's God's Number — *the worst case an all-knowing solver ever faces. The Rubik's Cube's state graph has about* 43 *quintillion vertices, and in* 2010 *a monumental computation pinned its God's Number at exactly* 20. The two-jug pouring puzzle, peg solitaire, and the sliding fifteen-puzzle live in state graphs the same way; only the size of the map changes.

SEEN BEFORE?

A **Chapter 3** invariant is exactly a proof that two positions lie in *different connected components* of the state graph — unreachability, made geographic. And parity arguments are the state graph's own checkerboard: every slide in the fifteen-puzzle swaps one tile with the blank, flipping the permutation's sign, so its state graph is *bipartite* — the same two-coloring logic as the chessboard, one level up.

Two more puzzles show how much the state graph carries once you learn to walk it. The first is a scandal a whole nation once staked money on; the second is a fractal hiding inside a toy.

WATCH THIS **#9 The fifteen-puzzle: why half the arrangements are ghosts**

In 1880 a puzzle craze swept America: fifteen numbered tiles in a 4 × 4 tray, one empty square, slide them back to order. The maker Sam Loyd offered $1,000 to anyone who could start from the tiles in order but with 14 and 15 swapped and slide them to perfect order. The prize was never claimed — and the state graph shows why it was safe money.

Make each of the 16! arrangements a vertex; join two when a single slide separates them. Now watch what *every* legal move does. A slide swaps the blank with a neighboring tile — and a swap of two things is a single transposition, which *flips the sign* of the permutation (Chapter 3). But the blank also moves one square on the board's checkerboard coloring, flipping *its* color. So the quantity

$$\text{(permutation parity)} \oplus \text{(blank's square color)}$$

never changes: it is an *invariant*, preserved by every edge of the state graph. Loyd's swap-only start differs from the solved position by one transposition — opposite parity — while the blank sits in the same corner, same color. The invariant disagrees. **The two positions live in *different connected components*: no path joins them, so no sequence of slides ever can.** The state graph of the fifteen-puzzle is not one map but *two islands of equal size* — exactly half of all arrangements are reachable, and the other half are ghosts you can see but never touch. Loyd was selling a walk between two islands with no bridge.

WATCH THIS

#10 Towers of Hanoi: a triangle that is secretly a fractal

Three pegs, three disks, move the stack one disk at a time, never a larger disk onto a smaller. Draw its state graph: a position is which peg each disk sits on, so there are $3^3 = 27$ vertices, and an edge joins positions one legal move apart. The picture that emerges is uncanny — it is the outline of a *Sierpiński triangle*, three corner-triangles joined at single bridge-vertices, the same self-similar shape you get by punching the middle out of a triangle forever. *The three corners of the big triangle are the three "all disks on one peg" positions. The shortest solution — the famous* $2^3 - 1 = 7$ *moves — is simply the straight path down one* edge *of the triangle, corner to corner.* Every fact about the puzzle is now a fact about the picture: the solution is optimal because the edge is the shortest route between corners; add a fourth disk and the whole triangle nests inside a bigger copy of itself, and the move count doubles-plus-one because each new level glues three copies together. A recursion you would otherwise prove by induction is sitting there as *geography*.

Peg solitaire, the two-jug pouring puzzle, sudoku, even a solved chess endgame: each is a state graph waiting to be walked. **The creative move is always the same: a puzzle you cannot see becomes a graph you can walk, and then reachability, shortest solutions, and impossibility are just questions about a map.**

Payoff: friends, strangers, and the inevitable triangle

A famous result that feels like it belongs at a party.

WONDER THIS

Six people are at a gathering. Any two of them are either friends or strangers (no in-between).

Prove that there must be either three mutual friends or three mutual strangers — a trio who all know each other, or a trio who are all unacquainted. With six people, one of these is unavoidable.

Model it as a graph: six vertices (people), and color every edge between two people *red* if they're friends, *blue* if strangers. We want to show there's always a monochromatic triangle — three vertices with all three connecting edges the same color. Here's the pigeonhole-powered argument (a callback to Chapter 2).

Fix any one person, P. They have five edges to the other five people, each red or blue. By pigeonhole, at least three of these five edges share a color — say (without loss of generality) three of them are red, connecting P to people A, B, C. Now look at the three edges *among* A, B, C. If *any* of them is red (say AB) then P, A, B form a red triangle, and we're done. But if *none* is red, then A, B, C are pairwise blue — a blue triangle! **Either way, a monochromatic triangle is forced**: there is no way to color the edges of six points in two colors and avoid one. *Pure structure guarantees the trio; the six people never had a choice in the matter.* This is the opening case of *Ramsey theory*, the deep principle that complete disorder is impossible — in any large enough structure, order must appear.

THE MOVE

The complement trick. Every graph G carries a *shadow*, its **complement** $\overline{G}$, which joins exactly the pairs G leaves apart. The six-people theorem is secretly about a graph and its shadow at once: let G be the *friendship* graph, so $\overline{G}$ is the *stranger* graph. A triangle in G is three mutual friends; a triangle in $\overline{G}$ is three mutual strangers — and the theorem says one of the two must appear. Whenever a problem carries two opposite relations — friend/stranger, joined/unjoined, chosen/left — draw G and its shadow $\overline{G}$, and "either this or that" becomes "a pattern in G, or the same pattern in $\overline{G}$."

We just saw that triangles are sometimes *unavoidable*. The mirror-image question (if you are *determined* to avoid triangles, how many edges can a graph have?) has an equally clean answer, but it sits just past the contest core, so we tuck it into an optional detour.

GOING FURTHER

optional • beyond AIME

Optional, beyond-contest. Skip this on a first pass — nothing later depends on it. It is here because it shows the chapter's tools (the handshake lemma plus a Chapter 11 inequality) combining into a polished extremal theorem.

The question. Build a graph on n vertices, forbidden ever to create a triangle, and pack in as many edges as you can. The honest way to dodge triangles is to split the vertices into two camps and join *only across* the divide: a bipartite graph has no odd cycle, hence no triangle. To maximise edges, make the camps as equal as possible — $\lfloor n/2 \rfloor$ on one side, $\lceil n/2 \rceil$ on the other, every cross-pair joined. That is the balanced complete bipartite graph $K_{\lfloor n/2 \rfloor, \lceil n/2 \rceil}$, with $\lfloor n/2 \rfloor \cdot \lceil n/2 \rceil = \lfloor n^2/4 \rfloor$ edges; for $n = 6$ that is $3 \times 3 = 9$.

Mantel's theorem says you can never beat it: a triangle-free graph on n vertices has at most $\lfloor n^2/4 \rfloor$ edges, achieved exactly by that balanced complete bipartite graph. Here is the degree-sum proof. Let the graph be triangle-free with n vertices and E edges, and take any edge uv. No vertex can be a neighbor of *both* u and v — a common neighbor w would make u, v, w a triangle — so the neighborhoods of u and v are disjoint among the n vertices, giving $\deg(u) + \deg(v) \le n$ for every edge uv. Now sum this across all E edges. The right-hand side becomes nE. On the left, each vertex v is counted once for every edge touching it, i.e. $\deg(v)$ times, so

$$\sum_{uv \in E} (\deg(u) + \deg(v)) = \sum_{v} \deg(v)^2 \le nE.$$

Finally apply the Chapter 11 power-mean (Cauchy–Schwarz) bound $\sum_v \deg(v)^2 \ge$

$\left(\sum_v \deg(v)\right)^2/n = (2E)^2/n$, where $\sum_v \deg(v) = 2E$ is the handshake lemma. Chaining,

$$\frac{4E^2}{n} \leq \sum_v \deg(v)^2 \leq nE \quad\Longrightarrow\quad E \leq \frac{n^2}{4},$$

and since E is an integer, $E \leq \lfloor n^2/4 \rfloor$. This bound is *sharp*: the balanced complete bipartite graph $K_{\lfloor n/2 \rfloor, \lceil n/2 \rceil}$ has exactly $\lfloor n^2/4 \rfloor$ edges and no triangle. When n is even, equality holds in the *real* inequality $E \leq n^2/4$, which does force every degree to equal $n/2$ and pins down $K_{n/2,n/2}$ uniquely; when n is odd the extremal graph is still $K_{\lfloor n/2 \rfloor, \lceil n/2 \rceil}$ — whose degrees are *not* all equal (for $n = 5$ they are $3, 3, 2, 2, 2$) — and proving it is the *only* extremum takes a little more care than we need here. An extremal graph problem dissolves into an algebra inequality you have already proved.

Where it leads. This is the triangle-free case of *Turán's theorem*, which caps the edges of any graph containing no K_{r+1}; the extremal graph is always the balanced complete r-partite graph. That result opens *extremal graph theory*, a whole field about how much structure a graph can carry before a forbidden pattern is forced.

The modeling gallery. Step back, and the whole chapter is one skill — *translate the problem into a graph* — practiced on seven different worlds:

Problem world	Graph model	Tool
a one-stroke drawing	vertices & edges	Euler trail (odd-degree count)
party friendships	people & friend-edges	handshake / degree count
a map or polyhedron	planar graph	Euler's formula $V - E + F = 2$
jobs & applicants	bipartite graph	Hall's condition
a domino board	grid graph	perfect matching
friends & strangers	2-colored K_6	Ramsey
a puzzle's positions	state graph	shortest path / distance

Don't wrestle the network — count one local feature and let it force the global verdict.

GOING FURTHER *optional • beyond AIME*

Optional, a bridge to linear algebra. Store a graph as its *adjacency matrix* A, with $A_{ij} = 1$ when vertices i and j are joined and 0 otherwise. Then matrix powers count *walks*: $(A^2)_{ij}$ is the number of two-step walks from i to j, and in general $(A^k)_{ij}$ counts k-step walks. A slick consequence: a triangle is a closed 3-walk, counted once from each of its three vertices in each of two directions, so the number of triangles in a graph is $\frac{1}{6}\operatorname{tr}(A^3)$. Turning "count the local structure" into matrix arithmetic is the doorway to *spectral graph theory*, where the *eigenvalues* of A quietly encode connectivity, colorability, and how well a network mixes.

What to try when you're stuck

When a problem involves networks, connections, routes, maps, or relationships:

1. **Model it as a graph.** Strip away everything but vertices (objects) and edges (connections).
2. **Count degrees and apply the handshake lemma** ($\sum \deg = 2E$; odd-degree vertices come in pairs).
3. **For one-stroke drawings,** count odd-degree vertices: a *connected* graph has an Euler trail iff exactly 0 or 2 are odd.
4. **For planar/map problems,** use Euler's formula $V - E + F = 2$ and its consequence $E \leq 3V - 6$ (simple graphs, $V \geq 3$) to bound edges or prove non-planarity.
5. **For bipartite or triangle-free planar graphs,** use the sharper bound $E \leq 2V - 4$ — this is what rules out $K_{3,3}$ (where $E \leq 3V - 6$ is too weak).
6. **For matching/assignment problems,** model as a bipartite graph and check Hall's condition (every group of k has $\geq k$ options).
7. **For boards, dominoes, and cell-visiting puzzles,** convert the board into its grid graph: cells become vertices, side-sharing cells become edges. Tilings are perfect matchings, tours are paths or cycles, and the checkerboard coloring is the bipartition.
8. **For puzzles with positions and moves** (river crossings, jug pouring, sliding tiles), build the *state graph*: positions are vertices, legal moves are edges. A solution is a path, the fewest moves is a graph distance, and an invariant is a proof that start and goal sit in different connected components.
9. **For "unavoidable patterns,"** combine pigeonhole with edge-coloring (Ramsey-style arguments).

THE MOVE

One-page recap. Graph theory abstracts the world into vertices and edges. Toolbox: **the handshake lemma** ($\sum \deg = 2E$, so odd-degree vertices are even in number), **Euler trails** (a *connected* graph is one-stroke-drawable iff 0 or 2 odd vertices), **Euler's formula** ($V - E + F = 2$ for connected planar graphs; its edge bound $E \leq 3V - 6$ for simple graphs with $V \geq 3$ rules out K_5, while the sharper bipartite bound $E \leq 2V - 4$ rules out $K_{3,3}$), **Hall's condition** (a bipartite perfect matching exists iff every group of k has $\geq k$ options), **the board-to-graph dictionary** (cells are vertices, shared sides are edges; a domino tiling is a perfect matching of the grid graph, whose bipartition is the checkerboard coloring), **the state graph** (a puzzle's positions are vertices, its legal moves are edges; solving is path-finding, and the fewest moves is a graph distance), and **Ramsey/pigeonhole** (six people force a monochromatic triangle). When a problem is about connections, draw the graph and count. (Three optional detours — *Mantel/Turán* on triangle-free edge counts, the *sufficiency* half of Hall, and the *Euler-versus-Hamilton* contrast — sit in the Going Further boxes for the curious.)

THE NAME FOR IT

Graph theory

A structure of *vertices* joined by *edges* is a *graph*. The identity $\sum \deg = 2E$ is the *handshake lemma;* a single-stroke tracing is an *Euler trail;* $V - E + F = 2$ for planar graphs is *Euler's formula;* the condition $|N(S)| \geq |S|$ governing bipartite matchings is *Hall's marriage theorem;* and the cap $\lfloor n^2/4 \rfloor$ on triangle-free edges is *Mantel's theorem* (the triangle-free case of *Turán's theorem*). Converting a board into its *grid graph* — cells as vertices, shared sides as edges — turns a domino tiling into a *perfect matching*. A tour visiting every vertex exactly once is a *Hamiltonian path* (returning home, a *Hamiltonian cycle*); the graph whose vertices are a puzzle's positions and whose edges are its legal moves is the puzzle's *state graph*. The guaranteed monochromatic triangle is the first case of *Ramsey theory*.

WHERE IT LEADS Graph theory scales into *network science*, *spectral graph theory* (eigenvalues of the adjacency matrix), and the *topology* of surfaces — Euler's $V - E + F = 2$ is the first *Euler characteristic*, a topological invariant. Hall's theorem opens *matching theory* and the wider study of combinatorial optimization and network flows; the edge bound $\lfloor n^2/4 \rfloor$ is the gateway to *extremal graph theory* (Turán's theorem and beyond); graph coloring grows into the *four-color theorem* and *chromatic* graph theory; and Ramsey theory becomes a vast field on the inevitability of order in large structures.

CURIOUS CARL • EUREKA

Carl: "Count one *local* thing — the degrees — and a *global* fact drops out. Fine. But now I'm greedy: which local count would tell me whether *everyone* on the left can be paired off, with nobody left stranded?"

The narrator: now you're hunting like a graph theorist, Carl. That exact question — matching a whole side with no lonely hearts left over — is one of the prettiest in the subject.

SAME BEAST, NEW COSTUME

The same "count locally, conclude globally" beast, in disguise across the chapter:

Handshake lemma → Euler's formula → Hall's condition → The mutilated chessboard → Ramsey $R(3,3) = 6$.

Each one checks a small local feature — a degree, a face-swap, a neighborhood size, a color count, an edge color — and forces a sweeping global verdict about the whole graph.

The practice ladder

THE PRACTICE LADDER *climb at your own pace*

LAUNCH *Warm the engine. Nudges below.*

1. Can the "envelope" figure (a rectangle with both diagonals drawn in and a triangular flap on top) be drawn without lifting the pen? Count its odd-degree vertices.

2. A graph has 6 vertices and 9 edges. What is the sum of all vertex degrees?

CORE BUILDER *Master the central moves.*

1. Prove that in any group of people, the number who have shaken an odd number of hands is even.

2. Verify Euler's formula $V - E + F = 2$ for the graph of an octahedron ($V = 6, E = 12$); find F.

3. Show that K_5 (five mutually connected vertices) cannot be drawn in the plane without edge crossings, using $E \le 3V - 6$.

4. Convert the $m \times n$ board into its grid graph. In terms of m and n, how many edges does the graph have? Check your formula on the 2×3 board by drawing it.

5. The standard soccer ball is stitched from 12 pentagons and 20 hexagons, with three seams meeting at every corner. Compute V, E, and F, and verify Euler's formula.

6. Draw the cube's graph in the plane and build its *dual*: drop a vertex inside each face and join two vertices whenever their faces share an edge. Which well-known graph do you get?

7. Decide whether $4, 4, 3, 1, 1, 1$ can be the degree sequence of a simple graph. Its sum is even, so the handshake lemma is satisfied — is that enough?

STRETCH *Transfer to unfamiliar ground.*

1. Prove that every tree with n vertices has exactly $n - 1$ edges, by induction on n (remove a leaf).

2. Prove the six-people theorem: any 2-coloring of the edges of K_6 contains a monochromatic triangle.

3. Four workers w_1, w_2, w_3, w_4 are to be assigned to four distinct jobs j_1, j_2, j_3, j_4. Their acceptable jobs are $w_1 : \{j_1, j_2\}$, $w_2 : \{j_2, j_3\}$, $w_3 : \{j_1, j_3\}$, $w_4 : \{j_1, j_2, j_3\}$. Drawing this as a bipartite graph, can all four be assigned to distinct acceptable jobs? Use Hall's condition to decide, and name the group (if any) that blocks a full matching.

4. A beetle crawls through every cell of a 5×5 board exactly once, always stepping to a side-adjacent cell. Prove that its start and end cells must both have the color of the corners. Then decide: can such a crawl begin at the cell directly beside a corner?

5. At a party of n married couples, nobody shakes hands with themselves or their own spouse. The host asks each of the other $2n - 1$ people how many hands they shook and receives $2n - 1$ *different* answers. Determine, with proof, how many hands the host's spouse shook.

6. Let G be the friendship graph of the six people above (an edge for each friendly pair) and let $\overline{G}$ be its *complement*, the stranger graph. Restate the six-people theorem as a claim about triangles in G and in $\overline{G}$.

THE FLASH *No clock — each cracks open in one flash of insight, once you find the angle. Nudges at the back.*

1. The Bridges of Königsberg have landmass degrees $3, 3, 3, 5$. Is a single-stroke bridge walk possible?

2. A connected planar graph has $V = 10$ and $F = 6$. How many edges does it have?

3. A double-nine domino set holds all 55 tiles with numbers 0–9. Can they be laid in one closed ring, matching ends touching?

PROOF LAB *Write it clean, in full sentences.*

1. Prove Euler's formula $V - E + F = 2$ for connected planar graphs by the cycle-stripping (demolition) argument, with a clear base case.

2. Prove that a connected graph has an Euler trail if and only if it has 0 or 2 vertices of odd degree, addressing both directions.

3. Prove the easy (necessary) direction of Hall's theorem in full: if some set S of people lists fewer than $|S|$ acceptable partners, then no perfect matching can exist. Identify exactly where the pigeonhole principle enters.

4. Prove that every finite simple planar graph is 6-colorable: its vertices can be colored with 6 colors so that no edge joins two vertices of the same color. (First show a planar graph always has a vertex of degree ≤ 5, using $E \leq 3V - 6$ and the handshake lemma; then induct, removing that vertex.)

5. Prove Gomory's theorem in full: if one black cell and one white cell are removed from the 8×8 board, the remaining 62 cells can always be tiled by dominoes. (Construct a closed side-step tour of all 64 cells; explain why the two removals cut it into arcs of even length, and why dominoes laid along an arc are legal.)

6. Prove that no *closed* knight's tour exists on the 5×5 board. (First show every knight move lands on the opposite checkerboard color, so the knight-move graph is bipartite; then ask what a closed tour of all 25 cells would be.)

REFLECTION *No nudge. Write a few sentences.*

1. For one problem above, what did abstracting to a graph let you ignore? Describe how stripping away detail made the structure visible.

NUDGES

Launch. (1) The envelope has exactly 2 odd-degree vertices — the two bottom corners, each of degree 3 (the top corners have degree 4 and the roof apex degree 2) — so yes: start at one bottom corner, end at the other. (2) $2 \times 9 = 18$. **Core.** (3) Handshake lemma: odd-degree vertices come in pairs. (4) $F = 2 - V + E = 2 - 6 + 12 = 8$. (5) K_5 has $\binom{5}{2} = 10$ edges but $3V - 6 = 9$; since $10 > 9$, it can't be planar. (6) Horizontal edges: $m(n-1)$; vertical edges: $n(m-1)$; total $2mn - m - n$. For the 2×3 board: $12 - 5 = 7$ edges. (7) $E = (5 \cdot 12 + 6 \cdot 20)/2 = 90$, $V = 2E/3 = 60$, $F = 32$; indeed $60 - 90 + 32 = 2$. (8) The *octahedron*. The cube has 6 faces, with 4 faces meeting around each corner and 3 faces at each vertex-of-the-dual, so the dual has 6 vertices each of degree 4 — the octahedron graph (cube and octahedron are dual solids). (9) No. The sum 14 is even, so the handshake test passes, but Havel–Hakimi refutes it: $4,4,3,1,1,1 \to 3,2,0,0,1 \to$ (sorted) $3,2,1,0,0$, and the leading 3 now forces a subtraction from a 0, producing a negative entry. (Directly: two degree-4 vertices among six would push each of the three degree-1 vertices above degree 1.) **Stretch.** (10) Removing a leaf reduces both V and E by 1; induct from the single vertex. (11) Fix a vertex; pigeonhole forces three same-colored edges; the triangle among their endpoints is monochromatic either way. (12) No — all four workers together accept only the jobs $\{j_1, j_2, j_3\}$, so the set S of all four has $|N(S)| = 3 < 4 = |S|$, violating Hall's condition; four workers cannot fill three jobs. (The blocking group is all four workers.) (13) The crawl is a Hamiltonian path in the grid graph, which alternates the two colors of the bipartition; with $25 = 13 + 12$ cells, both endpoints must lie in the 13-cell class — the corner color. A cell directly beside a corner has the minority color, so no crawl can begin there. (14) The host's spouse shook $n-1$ hands. The $2n-1$ answers are distinct values in $\{0,\dots,2n-2\}$, hence *all* of them. The $(2n-2)$-shaker shook everyone allowed, so every answerer except that person's spouse shook at least once — the 0-answerer must be that spouse. Remove this couple: everyone remaining shook the max-shaker and not the zero-shaker, so every count drops by exactly 1, and the same situation recurs with $n-1$ couples. Couples among the answerers thus pair off with counts summing to $2n-2$; the host's spouse, whose partner gave no answer, is left holding the self-paired middle value $n-1$. (15) G contains a triangle exactly when three of the six are mutual friends, and $\overline{G}$ contains one exactly when three are mutual strangers; the six-people theorem says at least one of $G, \overline{G}$ must contain a triangle — equivalently, $R(3,3) \le 6$. **The Flash.** (16) No — four odd-degree vertices exceed the limit of 2. (17) $E = V + F - 2 = 10 + 6 - 2 = 14$. (18) No — each of the ten numbers has degree $9 + 2 = 11$ (nine neighbors plus its double's loop), odd at all ten vertices; an Euler circuit tolerates zero odd vertices, a trail at most two. **Proof.** (19) Induct on the number of edges. If the connected graph contains a cycle, delete one edge of it: the graph stays connected, E drops by 1, and the two faces flanking that edge merge into one so F drops by 1, leaving $V - E + F$ unchanged. Repeat until no cycle remains — a spanning tree with $E = V - 1$ and $F = 1$ — where $V - (V-1) + 1 = 2$; so the alternating sum equalled 2 all along. (20) Necessary: as an Euler trail passes through a vertex it pairs an entering edge with a leaving edge, so every vertex has even degree except possibly the two endpoints; hence the number of odd-degree vertices is 0 (closed trail) or 2. Sufficient: given a connected graph with 0 or 2 odd vertices, if there are two, add a temporary edge joining them so all degrees become even; then Hierholzer's method builds an Euler circuit — follow unused edges until forced back to the start, then splice in detours from any vertex with edges still free until every edge is used. Deleting the temporary edge leaves the desired trail. (21) In a perfect matching each person of S is paired to a distinct partner, and every such partner is acceptable, hence lies in $N(S)$; so the matching injects S into $N(S)$, forcing $|N(S)| \ge |S|$. Contrapositive: if some S has $|N(S)| < |S|$, the $|S|$ people cannot be sent to *distinct* partners inside the smaller set $N(S)$ — by pigeonhole two would share a partner — so no perfect matching exists. The pigeonhole step is precisely the impossibility of an injection from a larger set into a smaller one. (22) A simple planar graph satisfies $E \le 3V - 6$, so by the handshake lemma $\sum \deg = 2E \le 6V - 12 < 6V$; the average degree is below 6, so some vertex v has $\deg(v) \le 5$ (otherwise $\sum \deg \ge 6V$). Induct on V: delete v, 6-color the remaining planar graph, then reinstate v; its at most 5 neighbors use at most 5 of the 6 colors, leaving one free for v. (23) Snake a single closed tour through all 64 cells (up one column, down the next, closing along the bottom row); consecutive cells on the tour share a side, and their colors strictly alternate black/white around the cycle. Removing one black and one white cell cuts the cycle into two arcs; because colors alternate, each arc runs between the two removed colors and so contains an even number of cells. Pair consecutive cells along each arc: every pair is edge-adjacent, hence a legal domino, and together they tile all 62 remaining cells. (24) Checkerboard-color the cells; a knight's move shifts the coordinates by $(\pm1, \pm2)$ or $(\pm2, \pm1)$, changing $i + j$ by an odd amount, so it always lands on the opposite color and the knight-move graph is bipartite. A closed knight's tour of all 25 cells would be a cycle through 25 vertices — an odd cycle — but a bipartite graph contains no odd cycle, since its two color classes must alternate around any cycle, forcing even length. Contradiction, so no closed tour exists.

CHAPTER EIGHTEEN

18

The Unpecked Chicken

The average outcome is often easy to find even when any single outcome is hopelessly complicated — if you add up the right small pieces.

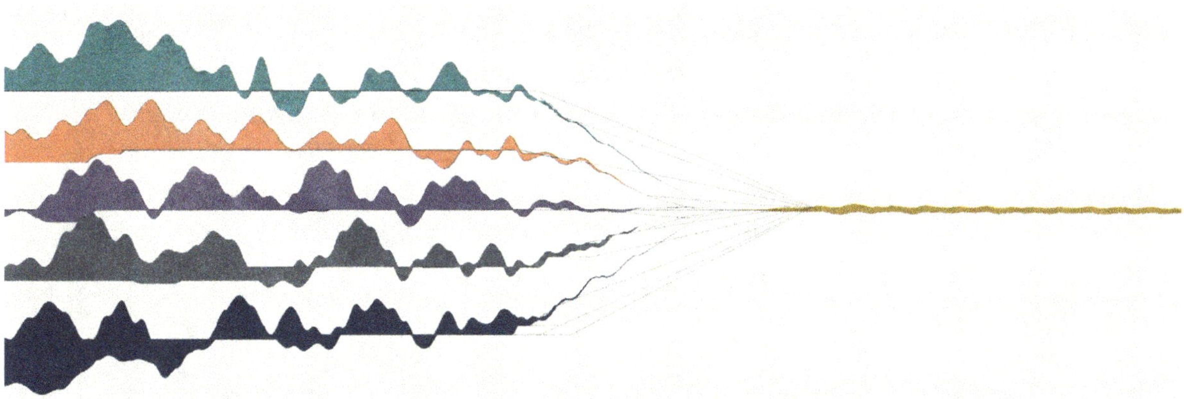

Expected value: add the small averages.

We open the final chapter with one of the most electric moments competition math has to offer: a problem that, under the lights of a national final with the clock running, looks utterly impossible, yet falls in a single line. It appeared on the MATHCOUNTS National Countdown Round, the buzzer-format showdown where the top students in the United States answer head-to-head in seconds.

WONDER THIS

One hundred chickens sit in a circle. At the same instant, *each* chicken randomly pecks one of its two immediate neighbors — left or right, each with probability $\frac{1}{2}$, all independently.

On average, how many chickens go un-pecked? A chicken's fate depends on *two* neighbors, who each depend on *their* neighbors — the whole circle is tangled together. How could you possibly count this in your head before the buzzer?

The instinct under pressure is panic: a hundred coupled coin flips wrapped around a circle, each chicken's outcome entangled with its neighbors'. Tracking the joint behavior is hopeless. But the technique this chapter is built on does not track joint behavior at all. It adds up one chicken at a time and lets the dependence sort itself out. Hold onto the chickens; in a few pages we will drop them in a single line. First, the idea that makes it trivial.

CURIOUS CARL • FIRST GUESS

Carl: "Expected number of people who get their own hat back? Right — I'll work out the whole probability distribution, every case of who-gets-whose, and *then* take the mean!"

The catch: put the giant table away, Carl — linearity lets you add each person's tiny $\frac{1}{n}$ chance and ignore the whole tangle. The average is just $n \cdot \frac{1}{n} = 1$, no distribution required.

WATCH THIS — #1 One chicken at a time

Slice it with indicators. For each chicken, let its indicator be 1 if it ends up un-pecked, 0 otherwise. The total number of un-pecked chickens is the sum of these 100 indicators, so its expected value is the sum of their individual probabilities — and by the circle's symmetry, every chicken has the *same* probability.

When is one chicken un-pecked? Focus on a single chicken C. It can only be pecked by its two neighbors. The chicken on C's left pecks C only if it pecks *right* (toward C); the chicken on C's right pecks C only if it pecks *left* (toward C). So C escapes entirely when *both* neighbors peck *away* from it:

$$P(\text{C un-pecked}) = \underbrace{P(\text{left neighbor pecks left})}_{1/2} \times \underbrace{P(\text{right neighbor pecks right})}_{1/2} = \frac{1}{4}.$$

These two neighbors make their choices independently, so the probabilities just multiply.

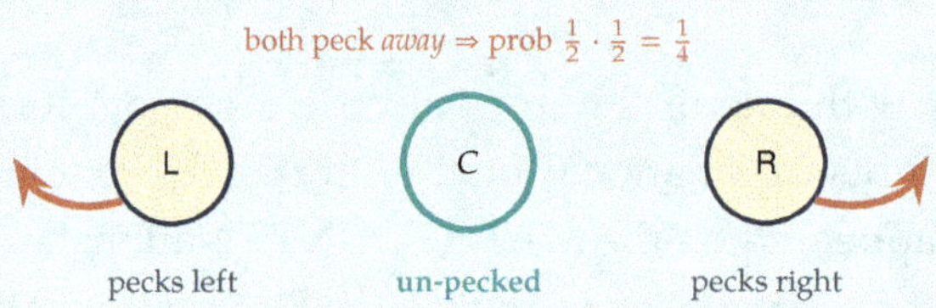

Add up the quarters. Each of the 100 chickens is un-pecked with probability $\frac{1}{4}$, so the expected number un-pecked is

$$100 \times \frac{1}{4} = \boxed{25}.$$

■

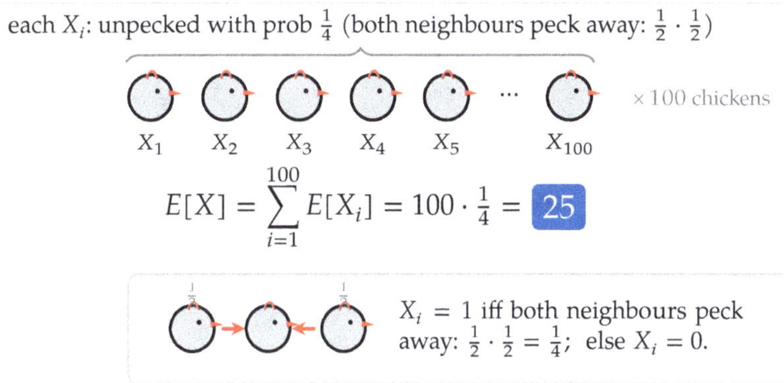

Linearity of expectation: each chicken is unpecked with probability $\frac{1}{4}$, so the expected number is $100 \times \frac{1}{4} = 25$.

A single multiplication and a single sum: that was all it took, no system of a hundred coupled variables ever written down. *The circle's tangle, which made the problem look terrifying, simply never mattered.* A competitor who sees the trick answers "25" almost instantly; a competitor who tries to model the whole circle never finishes. What did the work was a single principle, and it deserves a name.

THE TURN *from grind to insight*

The brute-force route is to build the full probability distribution of the un-pecked count — track all hundred coupled coin-flips around the circle, untangle how each chicken's fate hangs on its neighbors' neighbors, then take the mean of that monstrous distribution. The dependence wraps the whole ring together, and you never finish before the buzzer. **The creative turn** is to write the count as a sum of one-chicken *indicators*, find each easy probability ($\frac{1}{4}$), and add their expectations straight up. It wins because linearity of expectation lets you ignore every entanglement between the pieces — only the individual averages matter, and those are trivial.

THE MOVE

Expected value (the long-run average) is *linear*: the expected value of a sum is the sum of the expected values,

$$E[X_1 + X_2 + \cdots + X_n] = E[X_1] + E[X_2] + \cdots + E[X_n],$$

always — even when the X_i depend on each other in complicated ways. The strategy: write the complicated quantity you care about as a sum of simple *indicator* pieces (each 1 if some event happens, 0 if not), find each easy expectation, and add. Dependence doesn't matter; only the individual averages do.

That "dependence doesn't matter" clause is what made the chickens collapse, and it is what makes linearity miraculous in general. Computing the full probability distribution of a sum usually requires untangling how the parts interact. But linearity lets you ignore all of that and just add up the pieces. It turns intimidating averages into bookkeeping. The rest of this chapter sharpens the tool and points it at deck shuffles, coupon collecting, and a few more surprises.

Indicators: slicing a count into yes/no pieces

The workhorse technique is to express a count as a sum of *indicator variables*: each one a simple 0-or-1 that flags whether a particular event happened. The **expected value** of an indicator is just the probability of its event, and linearity glues them together.

WATCH THIS **#2 Everyone loses their hat**

Claim. n people check identical-looking hats, and the hats are returned in a random order. On average, how many people get their *own* hat back?

Slice into indicators. Let $X_i = 1$ if person i gets their own hat, and 0 otherwise. The total number of people with their own hat is $X = X_1 + X_2 + \cdots + X_n$. We want $E[X]$.

Each piece is easy. Person i is equally likely to receive any of the n hats, so the chance of getting their own is $\frac{1}{n}$. Thus $E[X_i] = \frac{1}{n}$ for every i.

Add them up. By linearity — ignoring entirely the tangled fact that whether one person gets their hat affects everyone else —

$$E[X] = \underbrace{\frac{1}{n} + \frac{1}{n} + \cdots + \frac{1}{n}}_{n \text{ terms}} = n \cdot \frac{1}{n} = 1.$$

So *on average exactly one person* gets their own hat back, no matter how large n is. ■

Trigger Any "expected number of..." question — matches, fixed points, collisions, successes — especially where the events overlap or depend on each other.

Do first Write the count as a sum of 0/1 indicators, compute each one's probability (its expected value), and add by linearity. Skip the joint distribution entirely.

Trapdoor Trying to compute the full distribution first. For an *average*, you never need it — linearity bypasses all the dependence.

Upgrade The same slicing handles expected matches, expected pairs, expected runs — any count is a sum of indicators, and indicators make dependence irrelevant.

That hat answer (exactly 1, always) is startling, and it showcases why linearity is beloved: the events are deeply dependent (if person 1 gets their hat, the others' chances shift), yet the average is immune to all of it. *The right move turned a hard distribution into n trivial probabilities.* Whenever you see "expected number of," your reflex should be: *slice it into indicators and add.*

△ TRAPDOOR

Linearity of expectation holds *always*, but the analogous rule for *variance* or for products does *not* — $E[XY] \neq E[X]E[Y]$ in general (only when the variables are independent). And linearity tells you the average, not the spread: knowing the expected number of own-hats is 1 says nothing about how often *nobody* gets their hat. Don't over-claim — linearity gives the mean and only the mean.

CURIOUS CARL • TRAPDOOR!

Emboldened, Carl grabs two indicators and writes $E[XY] = E[X]\,E[Y]$, then adds "and linearity needs them independent anyway, right?"

Narrator: Two slips at once, Carl. Linearity $E[X+Y] = E[X]+E[Y]$ *never* asks for independence — that is exactly its magic. But the *product* rule $E[XY] = E[X]E[Y]$ genuinely does need it; with two entangled indicators, that equation just breaks.

The showpiece: the coupon collector

Here's a problem that seems to demand heavy machinery, yet falls to linearity combined with one more idea: the expected wait for a repeated trial.

WONDER THIS

A cereal company hides one of n different toys in each box, uniformly at random. You want the *complete set*.

On average, how many boxes must you buy to collect all n toys? The first toy is easy, but the last few stragglers seem to take forever...

The genius is to break the quest into *stages* by how many distinct toys you already have, and to use the fact that waiting for an event of probability p takes, on average, $\frac{1}{p}$ trials.

WATCH THIS **#3 Collecting the whole set**

Set up the stages. Suppose you currently hold k distinct toys. The chance that the next box gives a *new* one is $\frac{n-k}{n}$ (the toys you don't yet own, over all toys). So the average number of boxes to advance from k toys to $k+1$ toys is the reciprocal,

$$\frac{n}{n-k}.$$

Add the stages with linearity. The total wait is the sum of the waits for each new toy, from $k = 0$ up to $k = n-1$:

$$E[\text{total}] = \sum_{k=0}^{n-1} \frac{n}{n-k} = n\left(\frac{1}{n} + \frac{1}{n-1} + \cdots + \frac{1}{1}\right) = n\,H_n,$$

where $H_n = 1 + \frac{1}{2} + \cdots + \frac{1}{n}$ is the n-th harmonic number.

$0 \to 1$ toy: $\frac{n}{n}$	$1 \to 2$: $\frac{n}{n-1}$	$\cdots$	**last toy:** $\frac{n}{1}$

sum $= n\,H_n$

The result. For $n = 6$ (say, six different toys), $E = 6H_6 = 6 \cdot \frac{49}{20} = 14.7$ boxes on average. Notice how the last toy dominates: the final stage alone averages n boxes, since a new toy is then rare. ■

The coupon collector is a fitting showpiece because it braids together three ideas from this book: *linearity* (sum the stage-waits), the *expected wait* $\frac{1}{p}$ for a repeated trial, and the *harmonic number* H_n that grew out of telescoping and sums in Chapter 6. The answer nH_n grows like $n \ln n$: since $H_n \approx \ln n$, the collector pays an extra logarithmic factor over n, and it's the last few stragglers that cost it, exactly as intuition warns. *One clean sum captures the whole frustrating experience of the near-complete set.*

TRY TO BREAK IT

Linearity of expectation never needs independence — but the product rule $E[XY] = E[X]E[Y]$ does, so cook up a case where the two verdicts split. Try a single fair coin with $X = 1$ for heads and $Y = 1$ for tails: they are as dependent as can be (never both 1), yet $E[X + Y] = \frac{1}{2} + \frac{1}{2} = 1$ holds perfectly, matching $E[X + Y] = 1$ since exactly one fires. Now test the product: XY is always 0, so $E[XY] = 0$, while $E[X]\,E[Y] = \frac{1}{2} \cdot \frac{1}{2} = \frac{1}{4}$ — the product rule shatters exactly where the sum rule sails through.

SEEN BEFORE?

The indicator-and-add method is **Chapter 1**'s double-counting wearing a probabilistic hat — counting expected contributions instead of objects. The harmonic number H_n is the telescoping/summation star of **Chapter 6**, and the "expected wait is $\frac{1}{p}$" fact is a geometric series (**Chapter 8**) in disguise. The whole book quietly converges here.

Back to the deck: the first Ace by symmetry

Linearity adds up indicators, but its cousin — plain symmetry — can hand you an average just as fast, with no indicators at all. Here is the classic.

WONDER THIS

Shuffle a standard 52-card deck thoroughly, then flip cards off the top one at a time until you reveal the first Ace.

On average, how many cards will you flip? The first Ace might show up immediately, or hide near the bottom. Over many shuffles, what's the *average* position of the first Ace?

A natural guess is "about a quarter of the way down," near card 13, since there are four Aces. The true answer is a little lower. Computing the probability that the first Ace lands in *exactly* position k, for every k, then summing k times those probabilities, is a genuine slog. Symmetry sidesteps all of it.

Think about the *gaps*. The four Aces divide the remaining 48 non-Ace cards into five groups: the cards before the first Ace, between the first and second, and so on, ending with the cards after the last Ace. By symmetry no gap is special, so on average those 48 cards split *evenly* among the five gaps: $\frac{48}{5} = 9.6$

cards before the first Ace. The first Ace is the very next card, at average position $9.6 + 1 = 10.6$. **On average you flip 10.6 cards to reach the first Ace.** The general formula is $\frac{n+1}{m+1} = \frac{53}{5}$ for m Aces among n cards, and the direct (if laborious) probability calculation agrees. The symmetry of the gaps gave us the average without ever computing a single position's probability, the same spirit as the chickens, where symmetry made every bird identical.

Payoff: a coin-flipping count, instantly

One last flourish, to show how linearity trivializes a question that looks like it needs real work.

WONDER THIS

Flip a fair coin 10 times. Consider every *pair* of flips (there are $\binom{10}{2} = 45$ pairs).

On average, how many pairs come up *both heads*? The pairs overlap and share flips — doesn't that make this hard?

The overlapping is exactly the dependence that linearity shrugs off. Define an indicator for each of the 45 pairs: 1 if both its flips are heads, 0 otherwise. Each pair is both-heads with probability $\frac{1}{2} \cdot \frac{1}{2} = \frac{1}{4}$, so each indicator has expected value $\frac{1}{4}$. By linearity, the expected number of both-heads pairs is simply

$$45 \cdot \frac{1}{4} = \frac{45}{4} = 11.25,$$

and the fact that the pairs share flips (that flip #3 appears in nine different pairs) never enters the calculation. **On average, 11.25 pairs are both heads**, obtained by adding 45 identical quarters. *Linearity doesn't care that the pieces are entangled; it just wants their averages, and adds.*

GOING FURTHER *optional • beyond AIME*

Optional — and a fitting last idea for the book. Expected value can do more than *measure* an average; it can *prove things exist*. The whole trick is one line: *if the expected number of "bad" events is less than 1, then some outcome must have zero bad events* — because an average below 1 is impossible if every single outcome had at least one. Nothing is ever constructed; the average alone forces a perfect specimen into existence.

A taste: color each edge of the complete graph K_n red or blue by an independent fair coin flip. A fixed set of 6 vertices spans $\binom{6}{2} = 15$ edges, and the chance those are *all one color* is $2 \cdot 2^{-15} = 2^{-14}$ — tiny. So the *expected* number of monochromatic 6-cliques is $\binom{n}{6} \cdot 2^{-14}$, and as soon as n is small enough for this to dip below 1, *some* coloring must have *no* monochromatic 6-clique at all — a graph proven to exist without anyone ever drawing it. This is Erdős's *probabilistic method*, and it turned expectation into one of the most powerful existence tools in modern mathematics. *Don't compute the distribution — add the tiny averages, and sometimes they hand you an object no one could build by hand.*

One more, even cleaner. Take any graph with m edges and split its vertices into two sides — a *cut* — counting the edges that cross. *Claim: some cut crosses at least $m/2$ edges.* Color each vertex red or blue by an independent fair coin. A given edge crosses exactly when its two endpoints land in different colors, which happens with probability $\frac{1}{2}$. Let X count the crossing edges; by linearity, $\mathbb{E}[X] = m \cdot \frac{1}{2} = m/2$, and — just as before — the endpoints of different edges may share vertices and be wildly entangled, but linearity never asks. Since a random cut *averages* $m/2$ crossings, some particular cut must *meet or beat* that average, so a cut of size at least $m/2$ exists. You never build the cut — you prove one must exist, from an average that no outcome can all fall below. *Expectation's final word in this book: an average, once computed, is a promise that something achieves it.*

What to try when you're stuck

When a problem asks for an average, an expected count, or "on average how many/how long":

1. **Reach for linearity of expectation.** Write the quantity as a sum of simpler random pieces.
2. **Use indicators.** For "expected number of...", make a 0/1 variable for each possible occurrence; its expectation is just the probability.
3. **Ignore dependence.** Linearity holds regardless — never compute the joint distribution for an average.
4. **For waiting times,** remember the expected wait for an event of probability p is $\frac{1}{p}$; sum these across stages.
5. **Exploit symmetry** (like the evenly-distributed gaps for the first Ace) to find an average without summing a distribution.

THE MOVE

One-page recap. Expected value is linear: $E[\sum X_i] = \sum E[X_i]$, always, dependence be damned. Toolbox: **indicators** (slice any count into 0/1 pieces; E[indicator] = probability), **linearity** (add the easy pieces, skip the joint distribution — as with the 100 pecking chickens, answer 25), **expected wait** $\frac{1}{p}$ (geometric trials, summed across stages for the coupon collector nH_n), and **symmetry** (evenly-shared gaps give the first-Ace average $\frac{n+1}{m+1}$). When you want an average, add up small pieces.

THE NAME FOR IT

Linearity of expectation

The long-run average of a random quantity is its *expected value*, and $E[X+Y] = E[X] + E[Y]$ regardless of dependence is the *linearity of expectation*. Slicing a count into 0/1 pieces uses *indicator random variables*, and the collection problem is the *coupon collector problem*.

WHERE IT LEADS Linearity of expectation is the workhorse of *probability theory* and the *probabilistic method*, which proves objects exist by showing a random one works — a technique that resolved long-standing problems across combinatorics and computer science. Expectation itself is the *Lebesgue integral* against a probability measure.

CURIOUS CARL • EUREKA

Carl: "Oh! I don't untangle the whole mess — I chop it into tiny yes-or-no pieces, find each little average, and add them up. Wait... that's just *double counting* again, isn't it? Chapter 1, wearing a probability hat."

SAME BEAST, NEW COSTUME

The same "add the small averages" beast, in disguise across the book:

Hat matches → Coupon collector → First-Ace by symmetry → Chicken pecks → Double counting.

Each one refuses to be tackled whole; slice it into indicator pieces, average each one alone, and let linearity ignore every entanglement between them.

You don't have to know each outcome to know the average — and with that, the book closes where it opened. Its last idea, sliced into indicator pieces, turns out to be double counting all along. The journey ends the way it began: not with more effort, but with a better way to look.

The practice ladder

THE PRACTICE LADDER *climb at your own pace*

LAUNCH *Warm the engine. Nudges below.*

1. What is the expected value of a single roll of a fair six-sided die?
2. In 100 fair coin flips, what is the expected number of heads? Use indicators.

CORE BUILDER *Master the central moves.*

1. Five people randomly receive their five hats back. What is the expected number who get their own hat?

2. Roll two dice. What is the expected value of their sum, by linearity?

3. In a random permutation of $1, 2, \ldots, n$, what is the expected number of fixed points (values left in their original position)?

STRETCH *Transfer to unfamiliar ground.*

1. A deck of 52 cards with 4 Aces is shuffled. Find the expected position of the first Ace using the gap-symmetry argument, and state the general $\frac{n+1}{m+1}$ formula.

2. There are $n = 6$ distinct coupons. Find the expected number of boxes to collect all six, and explain why the last coupon contributes the most.

3. n chickens sit in a circle and each pecks a random neighbor. Find the expected number of chickens pecked by *both* neighbors at once (pecked twice), and check that "un-pecked + pecked-once + pecked-twice" chickens average to n.

THE FLASH *No clock — each cracks open in one flash of insight, once you find the angle. Nudges at the back.*

1. Flip a fair coin until the first head. What is the expected number of flips?

2. Flip a fair coin 10 times. What is the expected number of pairs of flips that are both heads?

PROOF LAB *Write it clean, in full sentences.*

1. Prove linearity of expectation $E[X + Y] = E[X] + E[Y]$ from the definition, and explain why it requires no independence assumption.

2. Derive the coupon-collector expectation nH_n by summing the stage-waits $\frac{n}{n-k}$, justifying that the wait for a probability-p event averages $\frac{1}{p}$.

REFLECTION *No nudge. Write a few sentences.*

1. For one problem above, describe how slicing into indicators let you ignore the dependence between events. When did "don't compute the distribution" save you the most work?

NUDGES

Launch. (1) $\frac{1+2+3+4+5+6}{6} = 3.5$. (2) $100 \cdot \frac{1}{2} = 50$. **Core.** (3) Five indicators, each $\frac{1}{5}$; expected $= 1$. (4) Each die averages 3.5; sum = 7. (5) n indicators, each $\frac{1}{n}$; expected = 1. **Stretch.** (6) Gaps split 48 non-Aces evenly into five parts; first Ace at $\frac{48}{5} + 1 = 10.6 = \frac{53}{5}$; general $\frac{n+1}{m+1}$. (7) $6H_6 = 6 \cdot \frac{49}{20} = 14.7$; the last coupon's stage averages $n = 6$ boxes alone. (8) Pecked-twice means both neighbors peck inward, probability $\frac{1}{2} \cdot \frac{1}{2} = \frac{1}{4}$, so expected $\frac{n}{4}$; with un-pecked $\frac{n}{4}$ and pecked-once $\frac{n}{2}$, the three average to $\frac{n}{4} + \frac{n}{2} + \frac{n}{4} = n$, as they must. **The Flash.** (9) $\frac{1}{1/2} = 2$. (10) $\binom{10}{2} \cdot \frac{1}{4} = 45 \cdot \frac{1}{4} = 11.25$. **Proof.** (11) By definition $E[Z] = \sum_\omega Z(\omega)P(\omega)$ over the sample space, so $E[X+Y] = \sum_\omega (X(\omega)+Y(\omega))P(\omega) = \sum_\omega X(\omega)P(\omega) + \sum_\omega Y(\omega)P(\omega) = E[X]+E[Y]$. The split merely rearranges one sum against the fixed weights $P(\omega)$; the joint distribution of X and Y never enters, so no independence is needed. (12) A probability-p event on independent trials first occurs at a trial number that is geometric in p, with mean $\sum_{j\geq 1} j(1-p)^{j-1}p = \frac{1}{p}$. When k distinct toys are already held, a new box is fresh with probability $p_k = \frac{n-k}{n}$, so the wait for the next new toy averages $\frac{n}{n-k}$. By linearity the total average is $\sum_{k=0}^{n-1} \frac{n}{n-k} = n \sum_{j=1}^{n} \frac{1}{j} = nH_n$.

MATH LAB #4

Symmetry & Probability Casino

Two tables, one house rule: don't wrestle the whole tangle — average it. Count distinct objects by averaging what stays fixed; count long-run outcomes by averaging the pieces.

The last two chapters share a secret handshake. Burnside says: to count objects up to symmetry, *average* the number each symmetry leaves fixed. Linearity says: to find a long-run count, *average* the little pieces. Same house rule, two tables. Play both, then watch them shake hands at the Boss table.

CURIOUS CARL • FIRST GUESS

Carl: "To count the different necklaces, I'll list all of them and cross out the repeats by hand!"

The catch: the repeats aren't uniform, Carl — the all-one-color necklace has no rotated twins, so crossing-out miscounts. Don't list; *average the fixed-points*, and the uneven overcounting dissolves on its own.

STATION 1 • Burnside table: necklaces

Count the distinct 6-bead, 2-color *necklaces* under rotation (C_6). **Fill the fixed-point table** — for each of the 6 rotations, how many colorings does it leave unchanged? — then average.

House rule: a rotation fixes a coloring exactly when the coloring is constant on each of its cycles, so it fixes $2^{(\#\text{cycles})}$.

STATION 2 • Burnside table: the painted cube

How many distinct ways can you paint the six faces of a cube with 2 colors, counting two paintings the same if a rotation carries one to the other? Average over the cube's 24 rotations. (Group them: identity; face turns; edge turns; vertex turns — each type has its own cycle structure on the faces.)

STATION 3 • Expected-value table: quick bets

By *linearity* (no distributions!): (**a**) roll two fair dice — what is the expected sum? (**b**) roll one die ten times — what is the expected number of sixes? Write each quantity as a sum of simple pieces and add the averages.

STATION 4 • Expected-value table: the collector's wager

A cereal brand hides one of 6 equally likely toys per box. **On average, how many boxes must**

you buy to collect all six? Sum the stage-waits: the wait to get a *new* toy when you already hold k is geometric with success probability $\frac{6-k}{6}$.

STATION 5 • Boss — the two tables shake hands

Spin a random *labeled* 4-bead circular coloring — four fixed positions on a ring, each of the 16 colorings equally likely. **What is the expected number of the 4 rotations that leave your coloring unchanged?** Compute it as an *average over colorings*: $\frac{1}{|X|}\sum_{g\in G}|\mathrm{Fix}(g)|$, where $|X| = 16$ is the number of labeled colorings. The *same* numerator, divided instead by $|G| = 4$, is Burnside's orbit count. (Burnside and linearity, meeting in one sum.)

NUDGES

Peeks (try first!). **(1)** Fixed-counts for rotations by 0..5 beads: $2^6, 2^1, 2^2, 2^3, 2^2, 2^1 = 64, 2, 4, 8, 4, 2$; sum 84, so $84/6 = \boxed{14}$ necklaces. **(2)** Cube: $\frac{1}{24}(1\cdot 2^6 + 6\cdot 2^3 + 3\cdot 2^4 + 8\cdot 2^2 + 6\cdot 2^3) = \frac{240}{24} = \boxed{10}$. **(3a)** Each die averages 3.5; expected sum $= 3.5 + 3.5 = \boxed{7}$. **(3b)** Ten indicators, each $\frac{1}{6}$: expected sixes $= 10\cdot\frac{1}{6} = \frac{5}{3} \approx 1.67$. **(4)** $6H_6 = 6(1 + \frac{1}{2} + \cdots + \frac{1}{6}) = 6\cdot\frac{49}{20} = \boxed{14.7}$ boxes. **(5)** Fixed-counts for the 4 rotations: 16, 2, 4, 2; average $= \frac{16+2+4+2}{16} = \frac{24}{16} = \boxed{1.5}$. (And $\frac{1}{4}(16+2+4+2) = 6$ is the number of distinct necklaces — Burnside, hiding in an expectation.)

CURIOUS CARL • EUREKA

Carl: "Oh — both tables pay out the same way: don't fight the whole mess, just *average* the little parts and let the tangle cancel itself!"

The ideas are one idea

If you followed the SEEN BEFORE? boxes, you noticed that the eighteen chapters are not eighteen separate subjects. They are eighteen views of a small number of recurring moves. Here, gathered in one place, are the threads worth carrying away.

Double-counting is everywhere. The idea of Chapter 1 — count one thing two ways, set the answers equal — reappears as Pascal's identity (Chapter 7), as the proof of Burnside's lemma (Chapter 16), as the handshake lemma (Chapter 17), and as the indicator-and-add method for expectations (Chapter 18). Whenever two descriptions count the same set, an identity is born.

The Catalan numbers, three ways. They first appear by a reflection *bijection* (Chapter 4), return from a split-in-the-middle *recurrence* (Chapter 5), and finally fall out of a single self-referential equation $C = 1 + xC^2$ via *generating functions* (Chapter 8). Three methods, one sequence — the clearest proof that technique and object are different things.

Something that never changes. The invariants of Chapter 3 — a quantity pinned on one side of a line — are the same spirit as "a square is never negative" driving the inequalities of Chapter 11, and as Euler's $V - E + F = 2$ surviving every demolition move in Chapter 17. To prove something is impossible, find what it would have to change but can't.

Smallest things and descents. The "smallest counterexample" and monovariant arguments of Chapter 3 power the order-divides-φ reasoning of Chapter 10 and the cycle-stripping proof of Euler's formula in Chapter 17. Watch a quantity that can only decrease, and it must stop.

Remainders tame the infinite. The modular thinking of Chapter 9 fuses with the binomial theorem (Chapter 7) to build the roots-of-unity filter of Chapter 13, and underwrites Fermat, Euler, and Wilson in Chapter 10. When numbers grow astronomical, stop caring about them and start caring about their remainders.

Pigeonhole forces structure. The simple counting of Chapter 2 returns as the engine of the six-people Ramsey argument in Chapter 17: in any large enough structure, an unavoidable pattern must appear.

Optimization lives at equality and symmetry. The equality case of an inequality (Chapter 11) is the same instinct as the reflection that finds a shortest path (Chapter 14) and the "push toward balance" that maximizes a product. The extreme is almost always the symmetric point.

Impossible tools that still compute. Cardano's cubic formula (Chapter 12) reaches honest real roots only by detouring through $\sqrt{-1}$ — and that forced detour, not the equation $x^2 + 1 = 0$, is what made complex numbers (Chapter 13) impossible to dismiss. A tool earns its place the moment it is the only road to an answer you can already check.

The lesson beneath all of these: *a technique is worth learning not for the problems it solves but for the problems it reveals to be the same.* Carry the connections, not just the tricks.

The architecture of panic. When you sit down before a genuinely hard contest problem, the first feeling usually isn't inspiration — it is a quiet, creeping panic. The numbers look too big, the shapes too lopsided, the options too infinite. In those moments, remember that your job is not to *grind* but to *audit*. Don't ask "how do I calculate this?" Ask instead: *what is the hidden architecture here trying to protect?* If things are moving, look for what stays frozen (Chapter 3). If things are tangled, break them into lonely, independent pieces and add up their averages (Chapter 18). If the population is exploding, force it into narrow boxes (Chapter 2). If the configuration sprawls, walk to its edge and take the most extreme object (Chapter 3 again). The secret of creative mathematics is that the universe is remarkably lazy: it leans on the same handful of structural symmetries over and over. Your panic is just the fog before the landscape comes into view. Trust the tools, find the rhythm, and wait for the click.

The creative moves, in six families

Eighteen chapters, but far fewer *ideas*. Every method in this book is one way of *refusing the obvious fight*, and they fall into a handful of families — each a single instinct wearing many costumes. Here is the whole toolkit on one map. When a problem has you grinding, name the family it belongs to, and reach for its escape.

1. **Count it another way.** *The same tally, swept from a cleverer angle.*
DOUBLE COUNTING (Ch 1) — count the shadow, not the thing; THE BINOMIAL BRIDGE (Ch 7) — a coefficient *is* a count, so read the story; BURNSIDE (Ch 16) — don't divide by the symmetries, average what they fix; EXPECTED VALUE (Ch 18) — don't build the distribution, add the tiny averages.

2. **Compress the chaos.** *Squeeze too many objects into too few labels, and a collision — or a cycle — is forced.*
PIGEONHOLE (Ch 2) — design the boxes; MODULAR ARITHMETIC (Ch 9) — keep only the remainder, and choose the lens that exposes the obstruction; FERMAT & EULER (Ch 10) — don't compute the power, find its clock.

3. **Encode, shrink, collapse.** *Trade a hard object for an easy record, a big problem for a smaller copy, a long sum for its endpoints.*
BIJECTIONS (Ch 4) — a lossless file format for the object; RECURRENCES (Ch 5) — solve a smaller shadow, and choose the state well; TELESCOPING (Ch 6) — make the middle vanish; GENERATING FUNCTIONS (Ch 8) — hang the whole sequence on a polynomial.

4. **Preserve what can't change.** *Find the quantity a process cannot touch, and the impossible becomes obvious.*
PARITY, INVARIANTS, MONOVARIANTS, THE EXTREMAL PRINCIPLE (Ch 3) — watch the thing that is frozen, or can only move one way.

5. **Change the universe.** *Move the problem into a friendlier world, solve it there, and carry the answer home.*
THE CUBIC FORMULA (Ch 12) — don't complete the cube, split one unknown into two; COMPLEX NUMBERS (Ch 13) — don't fight the trig, step into the imaginary plane; GEOMETRY (Ch 14) — don't differentiate the radicals, draw the picture; MASS POINTS & AFFINE MAPS (Ch 15) — hang weights, or flatten the figure into a friendly shadow; GRAPHS (Ch 17) — don't wrestle the network, count one local feature.

6. **Exploit balance.** *The best configuration is almost always the most symmetric one.*
INEQUALITIES (Ch 11) — don't search for the optimum, read the equality case.

Six instincts — *count differently, compress, encode, preserve, transform, balance* — are the whole art. A technique is worth learning not for the one problem it cracks but for the family it belongs to: once you feel which instinct a problem is begging for, the particular method is almost a formality. That is the difference between a bag of tricks and a way of seeing — and it is the entire aim of this book.

CURIOUS CARL • EUREKA

Carl came into this book with three sharpened pencils, sure that being good at math meant grinding faster than everyone else. Eighteen chapters later, the pencils are mostly unused — and he finally says it out loud: these were never eighteen tricks to memorize. They were *six instincts*, each wearing a dozen costumes. He doesn't need to recall which trick unlocks which problem. He needs only to ask which instinct the problem is begging for, and the method very nearly picks itself.

It is, he thinks, exactly the sort of thing an eight-year-old with a slate once noticed — when, asked to add the numbers from 1 to 100, he refused to just start adding, looked again, and saw the pairs. Two thousand years before that, a man stepped into a bath, watched the water climb, and leapt up shouting the only word that ever really fits this feeling. Carl grins and says it too, quietly, for the whole book at once:

Eureka.

Glossary of tactic triggers

A field guide for the moment you're stuck. Match what you *see* in a problem to the move worth trying, and the chapter that develops it.

When you see...	Try...	Ch.
A quantity countable in two different ways	Double-count; set the counts equal	1
More items than categories	Pigeonhole: some box repeats	2
"Prove it's impossible / can never"	Find an invariant or parity	3
"Show this is unreachable"	A monovariant that only moves one way	3
"How many," hard to count directly	Bijection to an easy set	4
Identical items into labeled boxes	Stars and bars	4
Indexed by n, builds on smaller cases	Set up a recurrence	5
A long sum with a suspiciously clean answer	Telescoping: write each term as a difference	6
Powers of a sum; choose-k coefficients	Binomial theorem; cross the algebra/count bridge	7
"Ways to total n" from repeatable parts	Generating function; multiply series	8
A recurrence you want in closed form	Generating function; solve one equation	8
Last digits; huge powers; divisibility	Work modulo m; reduce early	9
Composite modulus	Split into primes; recombine (CRT)	9
Astronomical power mod m	Fermat/Euler: reduce the exponent	10
"Prove $A \geq B$" or find a max/min	AM-GM, Cauchy–Schwarz, or force a square	11
Optimizing a product with fixed sum	AM-GM; the optimum is all-equal	11
Cubic; a sum of two cube roots	Cardano: depress, then guess $t = u + v$	12
Rotations; regularly spaced points	Roots of unity; $z^n = 1$	13
$\cos n\theta$, $\sin n\theta$, powers of a rotation	De Moivre's theorem	13
Sieve every dth term of a sum	Roots-of-unity filter	13
Shortest bent path; a billiard bounce	Reflect a point, then go straight	14
Products of segments through a circle	Power of a point ($PT^2 = PA \cdot PB$)	14
Cevian or area ratios in a triangle	Mass points; area method; or affine squash	15
Count arrangements up to symmetry	Burnside: average the fixed-points	16
Networks, routes, connections	Model as a graph; count degrees	17
One-stroke drawing possible?	Count odd-degree vertices (0 or 2)	17
Planar / map / polyhedron counts	Euler's formula $V - E + F = 2$	17
"On average how many / how long"	Linearity of expectation; indicators	18
Expected wait for a repeated event	Mean wait is $1/p$; sum the stages	18

The deepest trigger of all isn't in this table: the feeling that a problem "rhymes" with one you've seen. When that happens, trust it — reach for the move that worked before, and check whether the same idea is hiding under new clothes.

Where to go next

This book is a doorway, not a warehouse. If a chapter lit something up, here is where to follow it — a short, opinionated shelf rather than an exhaustive list.

To keep solving problems. *The Art and Craft of Problem Solving*, Paul Zeitz — the warmest single book on how a solver actually thinks. *Problem-Solving Strategies*, Arthur Engel — a dense arsenal of olympiad techniques with hundreds of problems. *How to Solve It*, George Pólya — the classic on the *shape* of problem-solving, still unmatched. The *Art of Problem Solving* series (Volumes 1–2, and the online community) is the standard training ground for contest mathematics.

Counting and generating functions (Ch 1, 4–8). *Concrete Mathematics*, Graham, Knuth & Patashnik — the definitive, playful treatment of sums, recurrences and binomial coefficients. *generatingfunctionology*, Herbert Wilf — a short, free, brilliant book on the method of Chapter 8. *Enumerative Combinatorics*, Richard Stanley — the graduate reference, for the whole theory.

Number theory (Ch 9–10). *A Friendly Introduction to Number Theory*, Joseph Silverman — gentle and modern. *An Introduction to the Theory of Numbers*, Niven, Zuckerman & Montgomery — the thorough classic on Fermat, Euler and beyond.

Inequalities, geometry, graphs (Ch 11, 14–17). *The Cauchy–Schwarz Master Class*, J. Michael Steele — an elegant course built on a single inequality. *Geometry Revisited*, Coxeter & Greitzer — the beautiful classic behind Chapters 14–15. *Introduction to Graph Theory*, Douglas West — the standard, readable text for Chapter 17.

For the love of it. *Proofs from THE BOOK*, Aigner & Ziegler — a gallery of the most beautiful proofs in mathematics, in the exact spirit of this book. *Journey through Genius*, William Dunham — the great theorems told as history and drama. *The Man Who Loved Only Numbers*, Paul Hoffman — the life of Paul Erdős, who saw the whole subject as one connected game.

None of these is required, and none replaces the one thing that actually builds a mathematician: sitting with a problem longer than feels comfortable. Read widely, but solve first.

About the Author

Dr Hui Quan grew up in a small town, where a love of mathematics and physics took hold early and never let go. That start led to five years of rigorous training at the University of Science and Technology of China, and then to a PhD from Duke University — a path from a modest classroom to the frontier of research, walked one problem at a time.

For the last ten years he has taught fun mathematics classes to children, and it is that work, more than any other, that shaped this book. Watching young students discover a beautiful idea for themselves — the moment a hard figure suddenly turns simple — convinced him that mathematics is best taught the way it is actually found: through a hook, an experiment, and a theorem earned rather than announced. *The Art of Creative Math* is his attempt to put that experience on the page.

A wide reader, he returns most often to the *Odyssey* and the *Aeneid* — long journeys whose pleasure, like a good proof, is as much in the traveling as in the arrival. He enjoys every moment of teaching and learning mathematics, and hopes the reader will too.

Ascent Math Press publishes *The Art of Creative Math*. To learn about other volumes in the series, write to `hello@ascentmathpress.com`.

The Art of Creative Math

Eighteen Ideas That Make Hard Problems Easy

A companion volume of complete worked solutions is available separately.

Keep asking better questions.

ASCENT MATH PRESS

Index

Bold page numbers mark where a term is defined or developed.

www.ingramcontent.com/pod-product-compliance
Lightning Source LLC
LaVergne TN
LVHW080041210926
848481LV00011B/3251
9798906210067